MATBKGFT

Ideas of Space

Ideas of Space
Euclidean, Non-Euclidean, and Relativistic

JEREMY GRAY

Clarendon Press, Oxford
1979

Oxford University Press, Walton Street, Oxford OX2 6DP

OXFORD LONDON GLASGOW
NEW YORK TORONTO MELBOURNE WELLINGTON
IBADAN NAIROBI DAR ES SALAAM LUSAKA CAPE TOWN
KUALA LUMPUR SINGAPORE JAKARTA HONG KONG TOKYO
DELHI BOMBAY CALCUTTA MADRAS KARACHI

© Jeremy Gray 1979

All rights reserved. No part of this publication may be reproduced, stored in a retrieval system, or transmitted, in any form or by any means, electronic, mechanical, photocopying, recording, or otherwise, without the prior permission of Oxford University Press

British Library Cataloguing in Publication Data

Gray, Jeremy
  Ideas of Space
  1. Geometry—History
  2. Space and time—Mathematical models—History
  I. Title
  516   QA21   79–40401
  ISBN 0 19 853352 7

Typeset in 'Monotype' Times New Roman by
Eta Services (Typesetters) Ltd., Beccles, Suffolk
Printed in Great Britain by
Lowe & Brydone Printers Ltd., Thetford, Norfolk

Preface

I hope in this book to say something about mathematics, what it is and how it has been done. I shall discuss Greek and modern geometry, in particular what came to be known as the problem of parallels, that 'blot on geometry' as Saville* called it in 1621. The problem is this: if a line meets a vertical line obliquely, must it necessarily meet any horizontal line as well? Stated as simply as that it may sound trivial, but the charm of the problem is that although it can be stated in classical terms it cannot be solved without a dramatic change in fundamental ideas. Its resolution is elusive, difficult, and surprising. I shall pursue the matter further and discuss Einstein's theories of relativity, both special and general, and modern ideas of the shape of the universe.

The approach I have taken is largely historical and chronological. I have not avoided discussing difficult problems—indeed to have done so would be to have sacrificed my objective—but I have assumed no specialist mathematical knowledge. A working familiarity with simple equations and the elements of trigonometry, such as students of science and engineering possess, is all that is needed. It is my hope that the study of past insights into a problem provides as valid a way into mathematics as the polished answers we now seem to regard as best. By using the history we can analyse problems, exposing and discussing difficulties and confusions as they arise, and thus learn about mathematics itself, and in part this book is an attempt to understand mathematics as a dynamic activity. We shall often encounter connections between mathematics, philosophy, and truth, which run as subsidiary themes throughout. However, this is not strictly a history book. I have not hesitated to abandon the history when the thread of mathematics runs thin or turns aside from the main subject. The reader should not feel it necessary to read every word of the book, but he should select and skip as fancy suggests.

The book begins with early Greek mathematics, the Eastern legacy, and the transition to deductive and geometric thinking. Then we encounter parallels. The properties and problem of parallels were well formulated by the time of Euclid, and we start by looking at Greek and later Arab approaches. The second part of the book takes the story from Wallis, Saccheri, and Lambert to its resolution by Gauss, Lobachevskii, Bolyai, Riemann, and Beltrami. In contrast to most authors, who see the developments as primarily foundational, I see them as concerned more with the concepts and methods of geometry, and so I shall sketch the background of the nineteenth century theory of surfaces and the relevant analysis. Chapter 14

---

* Saville, H., *Thirteen lectures on the elements of Euclid*. Oxford University Press Oxford, 1621.

revisits the earlier material in the light of the later formulations; Chapter 15 summarizes the account and compares it with other versions. In the third part I give an account of Einstein's theories based on what has gone before, moving from a Newtonian–Euclidean picture to an Einsteinian–non-Euclidean one. This transition is often referred to in the literature, but rarely described. The book concludes with a brief modern account of gravitation, the nature of space, and black holes.

I believe that intelligible explanations of every subject can and should be made, giving their real flavour without descending to trivialities, and this has been my objective. I hope that this book will make mathematics accessible to some people who have been repelled by its technicalities, and that its historical approach will itself be of value to mathematicians.

It is with great pleasure that I thank the people who have helped me with this book: friends and colleagues who, by their interest and advice, have made it much better than it otherwise would have been. Among those who read it in whole or in part and commented valuably were Julia Annas, John Bell, David Charles, David Fowler, Luke Hodgkin, Clive Kilmister, Bill and Benita Parry, Colin Rourke, Ian Stewart, and Graeme Segal. I should also like to thank the reviewers whose comments so markedly helped to improve the book and of whom only Dana Scott is known to me by name, the British Society for the History of Mathematics for inviting me to address them on some of it and for the discussion afterwards, and Jennie Connell and the Open University typists whose excellent jobs of typing the manuscript helped to restore my confidence in it. Above all, I express my deepest thanks to my parents for their encouragement, comments, and advice.

Any helpful criticism and comments will be gratefully received. All mistakes that remain are mine.

Milton Keynes 1978   J.J.G.

Contents

Part 1

1 Early geometry   1
   The contrast between Greek and Babylonian mathematics raised. The limitations of rhetorical algebra. The earliest Greek mathematics, a search for deductive validity. Pythagoras' theorem known to the Babylonians, c. 1700 B.C.; a dissection proof based on the theory of figured numbers. Incommensurable magnitudes shown to exist; a conjectural account of their discovery. Reasons for the retention of geometry discussed. Exercises. Pythagorean triples, Greek and Babylonian methods. Plimpton 322. Theodorus on $\sqrt{3}$ to $\sqrt{17}$

2 Incommensurability   18
   The discovery of the irrationality of $\sqrt{2}$ led to further research into incommensurability, but not to a foundational crisis. Similarity and parallelism. Eudoxus on the nature of ratio. Exercises. The application of areas and 'geometric algebra'. Side and diameter numbers and the Euclidean algorithm

3 Euclidean geometry and the parallel postulate   29
   Geometry as the study of figures in space, and assumptions about space. Euclid's *Elements*. The parallel postulate; is such an assumption necessary? Consequences of the postulate; existence and uniqueness of parallels; equidistant lines. Various arguments to 'prove' the postulate, including those of Proclus, Aganis, and Nasir Eddin al-Tusi. Assumptions equivalent to the parallel postulate. Appendix on solid geometry and trigonometry. Exercises. A geometry on the sphere. Plane and spherical trigonometry

Part 2

4 Saccheri and his Western predecessors   51
   Revival of interest in the West in the 16th and 17th centuries. The work of Saccheri. Three hypotheses. HOA refuted. HAA discussed

5 J. H. Lambert's work   63
   Spherical geometry. The work of J. H. Lambert. Absolute nature of length. Angle sum and area. The imaginary sphere. Exercises

***6** Legendre's work*   69
    French lack of interest; except for Legendre. His 'refutations' of non-Euclidean geometries. Exercises

***7** Gauss's contribution*   74
    Kant. Gauss's work. Directed lines. A new definition of parallel. Corresponding points. The horocycle. Exercises

***8** Trigonometry*   82
    Trigonometric and hyperbolic functions. Exercises

***9** The first new geometries*   87
    Schweikart's Astral geometry. Taurinus' logarithmic–spherical geometry. Approximate agreement between the new geometries and the old. Appendix. Exercises

***10** The discoveries of Lobachevskii and Bolyai*   96
    The Bolyai's struggle. Absolute geometry. The work of Lobachevskii summarized and described. The prism theorem; the horocycle and horosphere; the projection map. Geometry on the horosphere is Euclidean; the fundamental formulae of hyperbolic geometry. Bolyai's work; squaring the circle. Summary. Priorities. Exercises. Appendix on spherical trigonometry, including Bolyai's proof of its absolute nature

***11** Curves and surfaces*   117
    Curves. Surfaces; co-ordinates; curvature. Intrinsic and extrinsic viewpoints. Geodesics. Minding's surface. Appendix on degeneracies

***12** Riemann on the foundations of geometry*   129
    Riemann's hypotheses. Co-ordinates on surfaces. Intrinsic geometry and curvature; metrical ideas at the basis of geometry

***13** Beltrami's ideas*   135
    Beltrami's model. Relative consistency of mathematics; foundational questions. Bolyai–Lobachevskii formulae

***14** New models and old arguments*   142
    Klein's model of elliptic geometry. Poincaré's conformal model of hyperbolic geometry. Revisiting the work of Wallis, Saccheri, and Legendre. Exercises

***15** Resumé*   155
    Summary of part II and other views

Part 3

***16** Non-Euclidean mechanics*   161
    Dostoevsky. Non-Euclidean mechanics

17 The question of absolute space   163
    Newton; Newtonian space. Relative motion. Magnetism and electricity. Ether drift?; absolute space? Einstein's idea. Kennedy–Thorndike experiment. The nature of space

18 Space, time, and space–time   176
    Space–time. Clocks and surveying. The invariance of distance. Change of axes. Invariance of the interval. Summary. Appendix on co-ordinate transformations. Exercises

19 Paradoxes of special relativity   190
    The 'paradoxes' of special relativity posed and solved

20 Gravitation and non-Euclidean geometry   196
    Gravity, its relative nature. The conventional element in measurement. The heated-plate and the cooled-plate universes, their connections with non-Euclidean geometries. The rubber-sheet model of gravity. Exercises

21 Speculations   204
    Gravitation in four-dimensional space–time, curvature, black holes. Speculations. Appendix, W. K. Clifford

22 Some last thoughts   212
    Meanings. Mathematical appendix on the connection between non-Euclidean geometry and special relativity, and on transformation groups

List of mathematicians and physicists   217

Bibliography   219

Index   223

Part 1

1 Early geometry

The civilizations of the Eastern Mediterranean and Middle East seem to have had an interest in mathematics from very early times. Egyptian and Babylonian scribes in about 1700 B.C. discussed not only matters of practical or commercial importance, but carried out abstract calculations as well. Estimates of areas and volumes are found alongside solutions to quite complicated numerical problems, and while the rules of mensuration are frequently wrong the skill with which the numerical problems were solved suggests very strongly that the Babylonians, at least, had a good grasp of elementary mathematics. The Babylonians, who generally surpassed the Egyptians, also developed an excellent positional astronomy which, it should be noted, had been preceded by over a thousand years of mathematics. However, the differences between Greek and Babylonian or Egyptian mathematics of around 300 B.C. are manifest. The Greeks were doing geometry, they were proving things, their methods were deductive, and there are signs of a lively interest in questions of rigour and logical validity. The Babylonians, on the other hand, had procedures but no proofs. Like the Greeks they possessed an impressive grasp of observational astronomy, but it did not rest on a theoretical or geometrical base. The Greeks, as is well known, gave mathematics a paramount position in their philosophical endeavours. Plato in numerous places directed his contemporaries towards mathematics. Aristotle drew many illustrations of argument from it, which are now collected in *Mathematics in Aristotle* by T. L. Heath (1949). At least one form of argument, that of *reductio ad absurdum*, was first used in mathematics before being used elsewhere. Naturally we look for the origins of this attitude to see how the transition to deductive mathematics might have been made.

Unfortunately the evidence for this period is scanty since Eudemus' *History* (*c.* 325 B.C.) is lost. Virtually the only nearly contemporary references to early Greek mathematics occur in Plato and Aristotle. Later writers, writing about work done three to eight hundred years before them, gave fuller accounts, but they brought to the task of a set of attitudes to mathematics which must have been different from those of their forerunners, and they may have credited the earlier mathematicians with a clarity and exactness of thought which they did not in fact possess. In some cases a later way of doing things has made it difficult to appreciate the problems originally raised; Zeno's paradoxes are a case in point. Furthermore, the transmission of the record may have been faulty. Happily, there are now several good histories of early mathematics available.

Foremost amongst the modern texts are *The evolution of the Euclidean elements* by W. R. Knorr (1975) and the many works of T. L. Heath, chiefly

2 Early geometry

his three-volume edition of the *Elements* (1956), his two volume *History of Greek mathematics* (1921), and the one-volume *Greek mathematics* (1930). Other texts are *Science awakening* by van der Waerden (1971), *Episodes from the early history of mathematics* by Asgar Aaboe (1964) and *The exact sciences in antiquity* by O. Neugebauer (1969). *A history of science* (two volumes) by Sarton (1959) takes a necessarily swifter but much broader view of the whole period, and there are many other discussions available, some of which are listed in the Bibliography. Individual topics are treated in the *Oxford classical dictionary* and the *Dictionary of scientific biography*. These studies have shed much light on our questions concerning the origin and evolution of deductive mathematics.

The spread of learning

There is a particular problem involved in the transmission of mathematics across a region or between cultures which is not found in the transmission of other ideas or techniques. Mathematics is not simply a collection of facts or 'results'; it is also a set of procedures for isolating problems and for solving them, a set of assumptions and permissible deductions, a way of thinking about things. Isolated from these habits of mind the individual results can not only seem trivial, but they can lose their specifically mathematical character and become observational or 'inductive' instead. Conversely, if the procedures are transmitted they act as a check upon the body of transmitted facts, allowing them to be re-derived or excluded if no proof can be found. Yet the compelling character of mathematics is to interest cultures in similar problems and so to drive them after similar information, even if they cannot understand each other's activity, so we should not necessarily assume that information has passed when we find two cultures doing similar things. Indeed the evidence of direct cultural contact between Greece and Mesopotamia is slight, consisting of a few opinions, like that of Herodotus,[1] who gave the gnomon and the division of the day into twelve hours a Babylonian origin. It is salutary to remember that he was wrong about the twelve-hour day, for Neugebauer[2] has established that it has an Egyptian origin.

The characteristics of Babylonian mathematics were a good number system and a rhetorical formulation of mathematical problems, which has led to their formulation of mathematics being called 'rhetorical algebra' by many writers, but the limitations of rhetorical algebra made its transmission difficult. Essentially, rhetorical algebra is a set of procedures expressed in words and illustrated with numerical examples for solving certain problems: finding solutions to equations, calculating areas and volumes. BM13901,[3] a tablet containing 24 similar problems, starts as follows:

---

[1] Herodotus, Book II, 336 109. Loeb edition, transl. A. D. Godley. Heinemann, London.
[2] Neugebauer (1969, p. 81).
[3] A picture of the tablet appears in Unit N4 of *The history of mathematics* (Open University Course AM289), p. 30, Open University, Milton Keynes; the unit contains a discussion of Babylonian mathematics.

I have added the area and the side of my square: 45. Take 1, divide it into two: 30, and multiply: 30 × 30 = 15. Add 15 and 45: 1, the square of 1. Subtract the 30 (which you had multiplied by itself) from the 1. You have 30, the side of the square.

Since all numbers have here been expressed as parts of 60, we should express the original equation as $x^2 + x = \frac{3}{4}$. The coefficient of x is 1; halve that and square it $(\frac{1}{2})^2 = \frac{1}{4}$. Add $\frac{1}{4}$ and $\frac{3}{4}$ (and form $x^2 + x + \frac{1}{4} = \frac{3}{4} + \frac{1}{4}$). Both sides are squares; take square roots $((x+\frac{1}{2})^2 = 1^2$. Therefore $x + \frac{1}{2} = 1$. Subtract the half from both sides; $x = \frac{1}{2}$.)

Now, a procedure expressed verbally is not a formula, it cannot be manipulated into equivalent forms or checked against another intended to solve the same problem. For these reasons rhetorical algebra is without proofs and can accommodate different and incompatible answers. It is tied to such operations with numbers as can be marshalled in words and therefore derived fairly directly from the elementary properties of number.

Teachers of it may have referred to a body of theory transmitted aurally which amplified the written remains we have, but it is most likely that the rhetorical techniques were taught as methods which check. If they were transmitted as such then they could well seem to anyone who encountered them a sterile body of facts without coherence or power to inform. In this form they probably did pass to the West, if only because of their use in commerce. We can trace the appearance of some rhetorical techniques in Greek mathematics, if not their passage there.[4]

There is only one way out of the profusion of contradictory and non-explanatory results in rhetorical algebra and that is to find a way of making coherent sense of its results—at least those which are right. I believe that it is in attempting to do that that the Greeks were led to geometry, not for its own sake but as a method of proof. The two go together and provide a deductive method for the treatment of mathematical problems. This point of view enables one to make sense of the otherwise confusing legends that have come down about the earliest Greek geometers, Thales and the school of Pythagoras.

Thales

According to Proclus[5] (A.D. 410–85) Thales (624?–548? B.C.) '... made many discoveries himself and taught the principles for many others to his successors, attacking some problems in a general way and others more empirically'. In particular he is supposed to have been the first to demonstrate that a circle is bisected by a diameter, that the base angles of an isosceles triangle are equal, that the angle in a semicircle is a right angle, and that two triangles are congruent[6] if they have two pairs of corresponding angles equal and the sides

---

[4] See p. 24.

[5] Proclus, *A commentary on the first book of Euclid's elements*, transl. G. R. Morrow, 1970, p. 52. I shall refer to this book as Proclus (Morrow) to distinguish it from the earlier English edition translated by Taylor.

[6] Two figures are *congruent* if they can be made to coincide exactly with one another.

between those angles equal. Proclus' commentary on the first of these results (Proclus (Morrow), p. 124) gives no clue as to Thales' proof, but Proclus did indicate how such a proof might go. Imagine that a diameter does not bisect the circle and then apply one part of the circle to the other by folding it over along the diameter. If the two parts are not to coincide one falls inside or outside the other somewhere, but then radial lines would not all be equal, which is absurd, and the result follows.

It is sometimes implied by some writers that mathematics is discovered in a way that reflects its logical order, so that propositions which appeal to others for their proof must likewise follow these others in their order of birth. Thus, to prove that the angle in a semicircle is a right angle[7] we nowadays use the result that the sum of the angles is two right angles. It is possible that Thales appealed to that result too, but we do not know. However, it is not possible to infer that, as a mathematician, he would automatically stop to prove the validity of any result he used. In the second heyday of foundational enquiry, the late nineteenth century, much of the work done was in providing theories whose conclusions were the basic assumptions of another man's work. When we are at the historical beginnings of deductive mathematics, therefore, we can well imagine that what was obvious to one man and not worthy of proof was an interesting puzzle to another. To extract basic assumptions would have taken time, as the deductive method was seen to apply to more and more of mathematics and to yield proofs of more already 'known' results. So it should not surprise us to see attributed to Thales results which seem to depend on others of which he is not known to have a proof and which he may have taken from the common store of factual knowledge. We cannot even be certain that he really knew what a proof was. Greek work on purely logical questions seems to have begun even later than their mathematical investigations.

Naïve geometry

We may reasonably imagine an initial, naïve formulation of mathematics in which numbers are represented by geometrical segments, say lines, squares, rectangles, or cubes. To represent a number[8] as a line one took a fixed, but arbitrary, unit length and repeated it as often as was necessary; representations of square numbers in terms of a unit square proceeded similarly. The method was traditional in Babylonian and Egyptian mathematics, and was referred to by Plato[9] as being common in Greek mathematics. The early, but not the late, work of the Pythagoreans was cast in such a form.

Our knowledge of Pythagoras is little better than our knowledge of Thales, and has been conveniently summarized by Kurt von Fritz.[10] No theorem can be reliably attributed to the man rather than his school, which seems to have

---

[7] This result is attributed to Thales by Pamphile in Diogenes Laertius (I, 24–5, p. 6, ed. Cobet), third century A.D.
[8] Number meant positive integer throughout this period.
[9] Knorr (1975, p. 172) cites Theaetetus, 148A, amongst other passages.
[10] *Dictionary of scientific biography* (1975), Vol. XI, pp. 219–25.

split into factions sometimes after the leader's death (c. 480 B.C.). In his day the school was primarily religious and philosophical in its preoccupations, and took as its programme the belief that all things are numbers. In mathematics their chief interest was in arithmetic, and they studied numbers geometrically through their representations as figured numbers. These are invariably described in histories of mathematics,[11] and will be briefly described here with a view to establishing a hypothetically Pythagorean proof of Pythagoras Theorem due to Bretschneider.[12] The result, that in a right-angled triangle ABC with a right angle at C, $AB^2 = AC^2 + CB^2$, was known to the Babylonians by about 1700 B.C. They used it over and over again in their problem solving, and one tablet, Plimpton 322, carries an impressive list of triples of numbers which reveals that they had a good grasp of how to construct integer triples a, b, c such that $c^2 = b^2 + a^2$ (see the Exercises for further details).

I have suggested that what is significant about this period is the move from procedures to proofs. One may therefore speak of theorems rather than results, a theorem being a result for which there is a proof. In this sense of 'theorem' the theorem which today bears the name of Pythagoras must surely, as Heath (1921, p. 145) suggested, have originated in the school, although we have no source allowing us to attribute it to Pythagoras directly and no proof attributable to the school has survived. Proclus, for instance, does not even attribute the result to Pythagoras.

I shall assume that the famous Theorem of Pythagoras originated in his school, which involves me in two further assertions:

(1) The result was known for *all* right-angled triangles and not just in various special cases (3, 4, 5; 5, 12, 13; ...).
(2) A *proof* of the theorem was also obtained by them in more or less the sense in which we understand 'proof' today.

My reason for believing this is that as a theorem rather than a conjecture the result is non-trivial. However, if the Pythagoreans lacked a proof of the theorem, it is difficult to see why its most immediate corollary, the existence of 'irrational numbers', would so disturb them, and we do have evidence suggesting that the discovery of irrationals shook them profoundly. How much easier it would have been to reject the conjecture and with it the fateful corollary.

Figured numbers and a proof of the Theorem of Pythagoras

A number may be represented by a row of uniformly spaced dots, and, since all numbers are built out of the unit by repetition, classical authors generally

---

[11] In addition to Knorr's book there is e.g. Sambursky's *The physical world of the Greeks* which gives a fuller account of the Pythagorean attitude to number.

[12] Knorr (1975, Chap. VI) gives it particular emphasis. For Bretschneider, see Heath (1956), note after Prop I, p. 47.

6 Early geometry

denied that 1 was a number, and numbers began at 2. 1 was, rather, the source of number. If the row can be broken into two equal rows the number is even, but if the dividing line hits a dot in the middle the number is odd. Figured numbers are obtained whenever the dots are arranged into shapes or figures (see Figs. 1.2 and 1.3). By no means every number can be figured in a pre-assigned way. The only numbers which can be represented as triangles are 1, 3, 6, 10, ...; the only squared numbers are, of course, 1, 4, 9, 16, ... The difference between two figured numbers or, more strictly, the number which when added to a figured number produces the next figured number of the same class is called a *gnomon*. Consecutive gnomons which generate the triangular numbers are just the numbers 1, 2, 3, 4, ...; consecutive gnomons for producing the squared numbers are precisely the odd numbers 3, 5, 7, ...

Rearranging the figures produces theorems about the decomposition of numbers. For instance, any square can be written as the sum of two consecu-

Fig. 1.1. (a) 8, an even number, divides into 4+4; (b) 7, an odd number, cannot be divided into two equal halves.

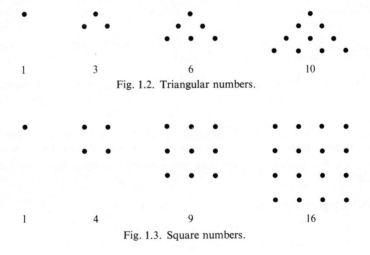

Fig. 1.2. Triangular numbers.

Fig. 1.3. Square numbers.

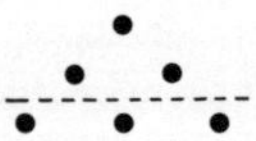

Fig. 1.4. A gnomon between two successive triangular numbers.

Fig. 1.5. A gnomon between two successive squares.

Fig. 1.6. A square written as the sum of two consecutive triangles.

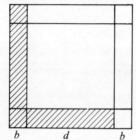
Fig. 1.7. $(d+2b)^2 = d^2 + 4b(d+b)$.

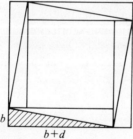
Fig. 1.8. The square on the hypotenuse is equal to the smaller square (d^2) plus two rectangles ($2b(b+d)$).

tive triangular numbers. In particular, this method of rearrangement can be made to yield a dissection proof of Pythagoras Theorem.

First observe that any square may be written as the sum of a smaller square plus four equal rectangles: take $d+2b$ as the side of the larger square, and then $(d+2b)^2 = d^2 + 4b(d+b)$ (Fig. 1.7). This can be arranged pictorially so that the four rectangles form two gnomons around the smaller one. Dividing these rectangles along their diagonals produces the suggestive figure shown next (Fig. 1.8) in which eight copies of a right-angled triangle appear. Notice

8 Early geometry

that this right-angled triangle can be specified in advance, since its shorter sides have lengths b and $b+d = a$, say, which can be chosen arbitrarily. We have that the square on the hypoteneuse of the triangle plus four times the triangle itself is equal to the larger square, but so, by a simple rearrangement, is the sum of the squares on the two smaller sides (Fig. 1.9). In words, the square on the hypoteneuse is equal to the sum of the squares on the other two sides.

This dissection proof is partly arithmetical and partly geometrical, since it partakes of both the dot and the line representations of numbers. Applied to triangles with sides all commensurable (i.e. for which a common measure exists so that each side represents an exact number of units) it yields the so-called Pythagorean triples: triples (a, b, c) such that $a^2 + b^2 = c^2$. Familiar examples of these are (3, 4, 5), (5, 12, 13), (6, 8, 10), ... One problem, solved in antiquity, was to find every such triple by means of a general formula. A solution is given in the Exercises. Babylonian interest in 'Pythagorean' number triples was marked, and whether or not there was any connection at all between the Mesopotamian and Greek mathematical schools, it should not surprise anyone that the Greeks also studied this fascinating topic (see the Exercises).

The simplest case of a right-angled triangle does not, however, lead to a number triple; the isosceles right-angled triangle formed by the diameter and two sides of a square cannot have its three sides all of integral length. By the theorem, if the side is one unit long the diameter must be $\sqrt{(1^2 + 1^2)} = \sqrt{2}$ units—we shall see that $\sqrt{2}$ must be irrational. The tablet YBC 7289 (Neugebauer 1969, plate 6a) gives a value for the diameter of 1·414 213 (if expressed in modern decimal notation). The modern approximation is 1·414 214..., but the Babylonian approximation is the closest that can be obtained to three *sexagesimal* places, the system of notation in which it is written. It is easy to see, by squaring, that the approximation is not correct, but did the Babylonians distinguish between the need to approximate $\frac{1}{7}$ as a sexagesimal or decimal fraction, and the need to approximate $\sqrt{2}$? The

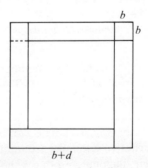

Fig. 1.9. By rearrangement the square on $b+d$ plus the square on b is also equal to the smaller square plus two rectangles.

fraction $\frac{1}{7}$ does not terminate because 7 does not divide the base 60 of the sexagesimal number system, but $\sqrt{2}$ is incommensurable with any rational multiple of the unit. We have no evidence with which to decide this question when it is applied to the Babylonians, but we can be quite certain that the Greeks not only perceived this distinction but made it and its consequences a prime object of research. Since in many ways the response of mathematicians in the fourth century B.C. to the discovery of incommensurability resembles that of modern mathematicians to the question of non-Euclidean geometry, it is worthwhile to look at this matter somewhat closely.

Root two

The famous irrationality of $\sqrt{2}$ can be demonstrated algebraically or geometrically. Algebraically, we argue this way, much as did Aristotle.[13] Suppose, for a contradiction, that $\sqrt{2}$ is rational. Then we can write $\sqrt{2} = p/q$, and by dividing if necessary ensure that p and q are not both even. Squaring both sides we obtain

$$2 = p^2/q^2$$

i.e. $2q^2 = p^2$. Now, 2 divides $2q^2$ and so must divide p^2 since the square of an odd number is odd, and the square of an even number is even, but if 2 divides p^2 it must divide p and so we obtain $p = 2r$, say. Therefore

$$2q^2 = p^2 = 4r^2$$

and therefore

$$q^2 = 2r^2.$$

However, the argument about 2 and p applies to 2 and q, so $q = 2s$, say, but then $\sqrt{2} = p/q = 2r/2s$ which contradicts our assumption that not both p and q are even. Therefore we see that we cannot write $\sqrt{2}$ as a rational at all. We would therefore say it is irrational. Yet $\sqrt{2}$ corresponds very clearly to a length—the hypoteneuse of an isosceles triangle or the diagonal of a square. If we try to compare the side and diagonal of a square we find that we cannot since the lengths are mutually incommensurable, and so any programme, such as the Pythagorean, which bases lengths naïvely upon the whole numbers is doomed to fail.

Knorr (1975, Chap. I)[14] has given a highly plausible account of how the Pythagorean proof might originally have looked using the figure encountered in Plato's *Meno* where Socrates instructs Meno's slave in geometry and hints at the incommensurability of side and diameter. The proof uses the fact that an even number squared is even but an odd number squared is odd, and runs

---

[13] Aristotle, *Prior analytics*, I 23, 41a, pp. 23–7.
[14] In Chapter III Knorr suggests that the discovery may have been made by Hippasus (*c.* 430 B.C.) after Parmenides and Zeno, who do not refer to it, but before Theodorus, who does, and Hippocrates, who has a good grasp of the main theorem.

10 Early geometry

as follows. Suppose for a contradiction that DB and DH are commensurable. If they are commensurable each can be viewed as representing a number, i.e. the number of times each is measured by their common measure. We can require that not both of these numbers are even. Then the squares DBHI and AGFE represent the squares of these numbers, and evidently AGFE is twice the size of the smaller square DBHI, so AGFE is even. So its side (which equals DH) is even, and so AGFE is divisible by 4. However, ABCD, which is one quarter of AGFE, also represents a number. Its double, which must be even, is the square DBHI. So both DBHI and AGFE are even and so both DB and DH are even, but we assumed that this was not the case. This is the contradiction we sought, our basic assumption is therefore untenable and incommensurability follows (by *reductio ad absurdum*).

This form of the proof, together with the defence which Knorr supplies of the requirement that a ratio can always be written in least terms, cannot, he points out, be how incommensurability was first discovered. The *reductio* method states the conclusion first, but the naïve programme of the early Pythagoreans could not admit the existence of incommensurable magnitudes. Accordingly, the discovery must have been made by first assigning numerical magnitudes to express the ratio and deducing from this the impossible conclusion that odd numbers are even. In order to avoid this contradiction the Pythagoreans could only break the argument at one point: the assumption of commensurability. The only alternative would be to abandon mathematics as a rational discipline, but by then too much effort and energy had been expended by the Pythagoreans for them to be willing to do this. One notices little inclination among mathematicians of the period to abandon their subject even when news of the discovery was made public outside the Pythagorean community.

Results obtained elsewhere in geometry may have further inclined people

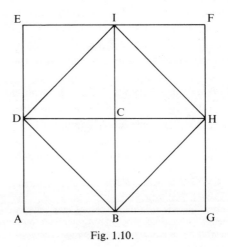

Fig. 1.10.

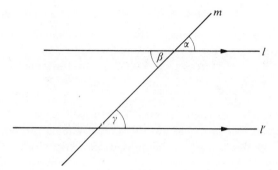

Fig. 1.11.

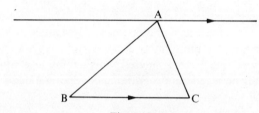

Fig. 1.12.

to persevere with mathematics. One example must suffice, and it will enable me to introduce parallel lines, fleetingly, into the story.

Two lines in a plane are *parallel* if they never meet. Essential properties of parallel lines, to be discussed in more detail later, include the following.

If l and l' are parallel lines both meeting a line m then the indicated angles are equal: $\alpha = \beta = \gamma$.

α and γ are said to be *corresponding* angles between parallel lines; β and γ are said to be *alternating* angles.

We are told by Proclus (Morrow edn., p. 298), who quotes Eudemus here, that the theorem that the angle sum of a triangle is $2R$ was first proved by Pythagoreans[15]; interestingly the proof is not the one given later in Euclid (I, 32). Instead, it takes a triangle ABC and adds the parallel to BC through A. By the above properties of parallels, taking $A\hat{B}C = \hat{B}$ and $A\hat{C}B = \hat{C}$, the angle sum of $\triangle$ ABC is

$$A\hat{B}C + B\hat{C}A + C\hat{A}B = \hat{B} + \hat{C} + C\hat{A}B = 2R$$

where $2R$ is the angle on a straight line.

This is a remarkable result for it establishes something non-trivial about triangles of any shape and thereby justifies the introduction of the concept of angle into mathematics, which it seems the Greeks were the first to do. As we shall see, the consequences of this theorem are quite far reaching.

[15] I shall denote a right angle by R, $90°$, or $\pi/2$ according to context.

12 Early geometry

Summary

The disadvantages of rhetorical algebra are that it is difficult to think in it for an extended period, that it is non-explanatory, and that it even contains contradictory estimates of areas and volumes. In response, the Greeks formulated geometry, and intended using it to attain proofs of propositions. The transition to the geometric method was part of the move towards a deductive scheme of knowledge. Against this background we can interpret the history of Thales and Pythagoras as part of such a systematization. Whether Thales did any of the things alleged of him we cannot know, but certainly the early workers could well stumble after the deductive method in more or less his fashion. It is not surprising that they should seek proofs of properties of diameters of circles or of angles in circles if what they are doing is slowly teasing out the deductive structure of a large inherited body of knowledge, some of it confusingly expressed or concealed in the analysis of other propositions. The angle in a semicircle was treated as a right angle in Sumerian times (see Coolidge 1963, p. 25), presumably as a descriptive fact about circles, and somebody must one day have pondered whether the angle in a semicircle had to be right. Given a general programme of seeking a deductive, philosophical ordering of mathematics it becomes a natural question to ask.

Furthermore, the behaviour of lines and circles, areas and angles is simple to analyse. It is simple enough for such a programme to have a chance of working and yet broad enough, encompassing all of elementary arithmetic, to be worthwhile. It throws up some quite surprising results, such as Pythagoras Theorem. Therefore we may imagine that the deductive/geometric programme progressed very rapidly once it began.

Certainly one appeal of geometry is that it treats existing things clearly, mathematically existing that is, but that is if anything better. Geometry then becomes an analysis of (true) reality, and the deductive method an enquiry into the world. It is a paradigm of the philosopher's quest for truth, easy enough to be a forcing ground in logic and comprehensive enough to be saying things. I cannot quite believe that the Greeks were interested in their triangles and circles just for their own sake, fascinating though they can be. If the theorems are, as I suggest, part of a programme which makes numerical work intelligible and which in its methods holds the promise of explaining things, then it is truly exciting. It raises the possibility that we might obtain deductive knowledge of the world, knowledge which starts from undeniable premises and by an irrefutable process yields a description of reality. We would then truly know the nature of things. It is an awe inspiring thought, truly a godlike thought, and one we have never wholly abandoned.[16]

[16] In this respect we do not differ greatly from the Greeks, for although it is frequently claimed that their science was insufficiently experimental and based too heavily on theory rather than observation or experiment there is less truth in this than is generally believed. This view, furthermore, rests on a false belief about modern science, that it is a matter of discovering the 'facts'. Every science proceeds from a theoretical position to a reasoned, or conditioned, view of the world; theory always governs the selection and interpretation of facts (see Part III, Chapter 17).

Early geometry

If that is the appeal of the deductive method, then in mathematics we must make sure that our premises are indeed undeniable and our deductions indisputable. No conclusion can be wholly accepted unless we can be sure that it has not been dishonestly implied in an earlier assumption—we must not beg the question. We might then test the scope of our geometry, to see just what sort of things it can do, and if it seems limited we might ask what natural extensions of our techniques can help us out. Such a programme was indeed carried out during the development of Greek mathematics up to and beyond Euclid.

Exercises

1.1 Pythagorean triples are triples of numbers (a, b, c) such that $a^2 + b^2 = c^2$, where a, b, and c are chosen from the numbers $1, 2, 3, \ldots$. Proclus (Morrow edn, p. 340) attributes to Pythagoras himself the following rule for finding such triples. Let n be an odd number; then the numbers $(n, (n^2-1)/2, (n^2+1)/2)$ form a Pythagorean triple. How might this rule be discovered via figured numbers? (Heath (1956, vol. I, p. 358) suggesting taking the gnomon between two successive squares in the case when it too is a square number.)

1.2 Proclus also gave a rule, which he attributed to Plato, and which starts with an even number m and produces the triple $(m, (m/2)^2 - 1, (m/2)^2 + 1)$. Check that this too is a Pythagorean triple. How might it be derived using figured numbers? (Knorr (1975, p. 156), following Iamblichus, suggests using a double gnomon in this form which jointly contains $4K = l^2$ dots.)

1.3 Find a Pythagorean triple which is not of either of the above forms.

1.4 Check that $(2pq, p^2-q^2, p^2+q^2)$ gives a Pythagorean triple for every choice of distinct p and q.

1.5 (Hard) Every Pythagorean triple is of the form given in Exercise 1.4—can you prove it?

Babylonian scribes wrote numbers on a sexagesimal system, for which Neugebauer has devised an elegant modern transcription. A semicolon separates the integral from the fractional part of any expression; commas divide each power of 60 from those on either side.

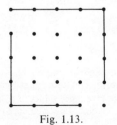

Fig. 1.13.

14 *Early geometry*

1.6 Check that

$$1; 24, 51, 10 = 1 + \frac{24}{60} + \frac{51}{60^2} + \frac{10}{60^3}$$

is the best sexagesimal approximation to $\sqrt{2}$ to three sexagesimal places.

1.7 The divisors of 60 are 2, 3, and 5. Call such numbers, and arbitrary products of them, *regular*. Check that only regular numbers have finite (terminating) sexagesimal reciprocals, and write some of them down, e.g. $\frac{1}{2} = 0; 30$ $\frac{1}{3} = 0; 20$.

1.8 Confirm the following entries in an extant Babylonian tablet of reciprocals (Neugebauer, 1969, p. 32):

$$\frac{1}{18} = 0; 3, 20, \qquad \frac{1}{36} = 0; 1, 40$$

$$\frac{1}{54} = 0; 1, 6, 40, \qquad \frac{1}{1, 21} = 0; 0, 44, 26, 40$$

1.9 The tablet Plimpton 322 in Columbia University contains a tablet remarkable for the insight that it shows the Babylonians of *c.* 1700 B.C. to have possessed into Pythagorean triples. The tablet survives with a break; the presence of glue along the break indicates that the damage was done since excavation (the other piece is lost). Of its four columns one numbers the rows 1 to 15, two others have headings which refer to sides of a triangle and which I shall call x, b, and d respectively (see Table 1.1).

Table 1.1 *Plimpton 322, with additions*

I	II (= b)	III (= d)	IV	h	p	q	p/q
[1, 59, 0], 15	1, 59	2, 49	1	2	12	5	12/5
[1, 56, 56], 58, 14, 50, 6, 15	56, 7	3, 12, 1	2	57, 36	64	27	64/27
[1, 55, 7], 41, 15, 33, 45	1, 16, 41	1, 50, 49	3	1, 20	75	32	75/32
[1], 5, [3, 1], 0, 29, 32, 52, 16	3, 31, 49	5, 9, 1	4	3, 45	125	54	125/54
[1], 48, 54, 1, 40	1, 5	1, 37	5	1, 12	9	4	9/4
[1], 47, 6, 41, 40	5, 19	8, 1	6	6	20	9	20/9
[1], 43, 11, 56, 28, 26, 40	38, 11	59, 1	7	45	54	25	54/25
[1], 41, 33, 59, 3, 45	13, 19	20, 49	8	16	32	15	32/15
[1], 38, 33, 36, 36	9, 1	12, 49	9	10	25	12	25/12
1, 35, 10, 2, 28, 27, 24, 26, 40	1, 22, 41	2, 16, 1	10	1, 48	81	40	81/40
1, 33, 45	45	1, 15	11	1	*	*	*
1, 29, 21, 54, 2, 15	27, 59	48, 49	12	40	48	25	48/25
[1], 27, 0, 3, 45	7, 12, 1	4, 49	13	4	15	8	15/8
1, 25, 48, 51, 35, 6, 40	29, 31	53, 49	14	45	50	27	50/27
[1], 23, 13, 46, 40	56	53	15	1, 30	9	5	9/5

Neugebauer's (1969, p. 37) transcription of Plimpton 322, with his corrected entries underlined. The zero ciphers are conjectural. The last four columns are additions.

* No values of p and q exist which give these values of b, d, and h; $p = 2$, $q = 1$ gives the similar triangle $b = 3$, $d = 5$, $h = 4$ which the scribe may have identified with the triangle $b = 3/4$, $d = 5/4$, $h = 1$.

Early geometry   15

Check that the entries are derived according to the rule $b = p^2 - q^2$, $h = 2pq$, $d = p^2 + q^2$, and $x = d/h$, and that they are arranged in descending values of x. Check that p and q are always regular numbers and suggest a reason for this by writing d/h in terms of p and q. Check that if p and q are restricted to lie in the ranges $1 < p < 180$, $1 < q < 60$ and $\sqrt{3} < p/q < 1 + \sqrt{2}$ every value of b, h, d occurs except one. Find this value.

1.10 Theodorus is quoted by Plato (Theaetetus 147C–148B) as having proved $\sqrt{3}, \sqrt{5}, \sqrt{6}$ up to $\sqrt{17}$ irrational (if I may supply the modern word), but at $\sqrt{17}$ he encountered a difficulty and stopped. The following exercises are taken from a series of theorems given by Knorr (1975) and are designed to show how theorems in figured numbers could have been used by Theodorus to obtain his results. Throughout, (a, b, c) denotes a Pythagorean triple. Try to prove them using arrangements of dots.

(a) Show that if two of a, b, and c are even, so is the third.
(b) Show that if one term in the triple is odd then so is a second term, and the third is even.
(c) Show that if c is even all three terms are even, but that if one term is odd then c is odd.
(d) If c is divisible by 4 then so are a and b.
(e) If not all of a, b, c are divislble by 4 then either a or b is, and it is the only term so divisible.

1.11 I continue the exercises above, with reference to Theodorus' results. When n is odd, $\sqrt{n}$ is exhibited as one side of a triangle with hypotenuse $(n+1)/2$ and other side $(n-1)/2$. When n is even, $\sqrt{n}$ is constructed to lie as half a side in the triangle with sides $(2\sqrt{n}, n-1, n+1)$.

(a) Show that $\sqrt{3}$ is irrational by considering the triangle shown in Fig. 1.14. Thus the hypotenuse is even. Take a and b in least terms and apply Exercise 1.10(c) to obtain a contradiction.
(b) Generalize (a) above to obtain a theorem about numbers n of the form $4K+3$. (Why is $4K+3$ never a square?)
(c) $\sqrt{5}$ occurs in the triangle in Fig. 1.15a or, in least terms, in the triangle in Fig. 1.15(b) if it is to be rational. Deduce that $\sqrt{5}$ is irrational.

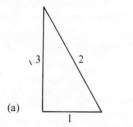
(a)

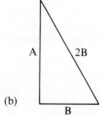
(b)

Fig. 1.14.

16 Early geometry

Fig. 1.15.

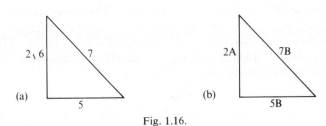

Fig. 1.16.

(d) Generalize to numbers of the form $8n+5$.
(e) List the numbers n for which you have shown $\sqrt{n}$ to be irrational and n less than 17.
(f) Show that $\sqrt{6}$ is irrational by considering the triangle in Fig. 1.16.
(g) Generalize to numbers of the form $2(2K+1)$.
(h) Show that the methods so far established break down at $\sqrt{17}$.
(i) Show by example that some numbers of the form $8n+1$ are square.

There are several problems which an active programme in elementary geometry would discuss. These include the famous straight-edge and circle constructions, the more general problems of constructing a regular figure of n sides, and ultimately constructing an arbitrary segment and an arbitrary angle. Constructing angles is harder than constructing segments, since segments can be arbitrarily divided but angles cannot. Indeed, it is in general impossible even to trisect an angle. All these elementary problems were indeed discussed in classical times, but of course not 'solved' then. In fact, as elementary problems they are unsolvable. However, some were reinterpreted in higher geometry and solved there, the distinction being that three-dimensional methods and conic sections were used. Greek geometers, according to Pappus,[17] classified problems as plane, solid, or linear meaning straight edge and circle only, three dimensional, or involving *ad hoc* curves, of which they had maybe a dozen. (It seems that the study of conic sections begins with their use, by Menaechmus, to solve the problem of duplicating the cube.) However, it remains a good mathematical question, Pappus said, as to whether

---

[17] Pappus, *Collectio*, Book III; see Heath, 1921, vol. I, p. 218, vol. II, p. 362.

a solution in, say, linear terms cannot be reduced to a more elementary one in solid or preferably plane terms.

1.12 Find out what you can about the resolution of these problems in classical times, say by consulting the works of Heath or van der Waerden. You should observe that mathematicians did not wait for a resolution of the problem of irrationals before going about their business. What does this tell you about mathematics as a historical or a logical activity?

2 Incommensurability

In some ways the discovery of the irrationals may have been the first good piece of pure mathematics. Only a mathematician could be detained by it, since the approximations to irrationals are easy to obtain and its importance rests upon the effect it has on an already sophisticated programme. The discovery put the naïve views of the early Pythagoreans into a state of self-contradiction, but it did not provoke a foundational crisis, in the sense that mathematicians devoted themselves to seeking new foundations for their subject.[1] Instead, there was a steady elaboration of new mathematics, in particular a thorough investigation of incommensurability. The problem is not what, conceivably, could a magnitude incommensurable with the unit be, but rather how many different incommensurables are there? It was shown by Theodorus that the magnitudes we would write as $\sqrt{3}$, $\sqrt{5}$, $\sqrt{6}$ and so on up to $\sqrt{17}$ were all incommensurable with 1, and by Theaetetus that $\sqrt{n}$ is incommensurable with 1 whenever n is an integer not already a square (obviously if $n = m^2$, $\sqrt{n} = m$). Furthermore, these different magnitudes are frequently incommensurable with each other: $\sqrt{2}$ and $\sqrt{3}$ are, for instance, although $\sqrt{2}$ and $\sqrt{8} = 2\sqrt{2}$ obviously are not. Compounded magnitudes, such as $\sqrt{5} + 2\sqrt{5}$ provide examples of still more mutually incommensurable quantities. Greek mathematicians distinguished the first class of magnitudes from the second for, although $\sqrt{2}$ is incommensurable with 1, its square, 2, is not. Lines of this kind were called commensurable in square and, unlike our terminology to-day, rational, but I shall only use rational in its modern sense and regard it as synonymous with 'commensurable with the unit'.

It is sometimes suggested that the discovery of incommensurability forced the Greeks to abandon algebra (where such a concept was unthinkable) and turn to geometry (where it can be formulated without contradiction). Proponents of this view speak of a geometrization of algebra, and regard much of Euclid's *Elements* (e.g. Books II and VII–IV) as a geometric algebra,[2] in which the subject matter is somehow algebra but the methods are geometrical. Van der Waerden (1971) writes that the Greeks must have had another reason apart from delight in the visible for turning to the geometrization of algebra:

... indeed this is not hard to find: it is the discovery of the irrational, which, as Pappus tells us, actually originated in the Pythagorean school. $\sqrt{2}$ is intelligible as a

---

[1] The first 200 years of development of the calculus, from Newton and Leibnitz to Weierstrass or Dedekind took place under a similar uncertainty.

[2] The term was introduced by H. G. Zeuthen in his study of conic sections: *Die Lehre von den Kegelschnitten in Altertum*. Copenhagen (1886).

segment. Geometrical algebra is valid also for irrational segments and is nevertheless an exact science. It is, therefore, a logical necessity, not the mere delight in the visible which compelled the Pythagoreans to transmit their algebra into geometric form.

There is no logical necessity about it. It would be quite possible to persevere with an arithmetic of natural numbers to which was adjoined such new quantities as, say, arose in the solution of equations. There is nothing more intelligible about a geometric segment than a root of an equation, unless you have already acquired a geometric habit of thought. Rather than turning from algebra to geometry, I suggest that the Greeks were already commited to geometry. At first they made a naïve identification of geometry with arithmetic for the purposes of using geometrical techniques to discover, prove, and systematize results about numbers. The discovery of the irrationals in the context of geometry inclined them to persist with geometry but eroded their faith in the naïve arithmetical formulation of magnitude. In the more general setting of philosophy the Pythagorean view was corporealist, identifying things in the intellect from things in the world. This became modified, and an idealism developed which was typical of mathematical philosophy and powerfully stated by Plato, in which things in the intellect are the more truly real. Aristotle deplored the Pythagorean influence on his mentor,[3] and took a realist position, viewing mathematical properties as abstracted from real objects (see Körner 1971, Chap. I).

Contemporary with the research into irrationals was an investigation of the concepts of similarity and parallelism, two concepts with which we shall be much concerned in the rest of this book. Two figures are *similar* if they 'have their angles severally equal and the sides about the equal angles proportional' (Euclid *Elements* VI, definition 1) where one must add, with Heath, that the corresponding sides must be opposite to equal angles. Two similar figures have, therefore, the same shape but not necessarily the same size. Two lines in a plane are said to be *parallel* if they never meet, however, for they are extended (a paraphrase of Euclid, *Elements* I, definition 23).

We have already seen that the Pythagoreans introduced parallel lines to prove that the angle sum of any triangle is $2R$. There is an intimate connection between similarity and parallelism; if a line *l* parallel to the base BC of a triangle ABC meets AB at B' and AC at C' (see Fig. 2.1) then B'C':BC = AB':AB = AC':AC, the angles $\widehat{AB'C'}$ and $\widehat{ABC}$ are equal, as are $\widehat{AC'B'}$ and $\widehat{ACB}$, and the triangles AB'C' and ABC are similar. Conversely, if in the same figure it is known instead that the triangles AB'C' and ACB are similar then it follows that B'C' is parallel to BC.

Similarity is a richer concept than congruence or equality of figures, and seems to have been studied very early. Hippocrates of Chios (*c*. 430 B.C.) made masterly use of it in his work, particularly the result that if ABC and A'B'C' are two similar figures the ratio of their areas equals the ratio of the squares on the corresponding sides, i.e. area ABC:area A'B'C' = $AB^2:A'B'^2$.

---

[3] See D. R. Dicks' argument for the Pythagorean influence (Dicks 1970, p. 63).

20 Incommensurability

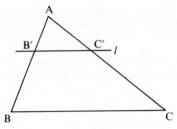

Fig. 2.1.

It also enables one to give a quick proof of Pythagoras' Theorem, which Heath (1956, vol. I, p. 353) suggests may have been how the theorem was originally proved.

Pythagoras theorem

Let ABC be a triangle right angled at B, and let the angles at A and C be α and γ respectively, so that $\alpha+\gamma = 90°$. Draw the perpendicular from B to AC and let it meet AC at D. Then BD divides ABC into two angles ABD and DBC, and we have

$$A\hat{B}D + D\hat{B}C = A\hat{B}C = 90°. \tag{1}$$

However, $A\hat{D}B = 90°$, so in $\triangle$ ADB

$$\alpha + A\hat{B}D = 90°$$

and, since we already know that $\alpha+\gamma = 90°$

$$A\hat{B}D = \gamma$$

and, in the same way

$$D\hat{B}C = \alpha.$$

Therefore the three right-angled triangles ABC, ADB, BDC have angles α, γ, and $90°$ and so are similar. From the similarity of triangles ABC and ADB we deduce that

$$\frac{AD}{AB} = \frac{AC-DC}{AB} = \frac{AB}{AC}$$

i.e.

$$AC^2 = AB^2 + AC \cdot DC$$

and from △ABC and △BDC we deduce that

$$\frac{DC}{BC} = \frac{BC}{AC} \tag{a}$$

i.e.

AC.DC = BC²

Accordingly $\underline{AC^2 = AB^2 + BC^2}$—the famous formula.

Step (a) is not trivial, however, and in proving that we can cross multiply, i.e.

$$\frac{BC}{AC} = \frac{DC}{BC} \quad \text{implies} \quad BC^2 = AC.DC$$

an argument is necessary which you may try to supply if you wish. You must show that

$$\frac{AB}{EF} = \frac{EH}{AD} \quad \text{implies} \quad AB.AC = EF.EH$$

for rectangles ABCD and EFGH. Euclid's proof can be found in *Elements*, VI, 16.

Now, however much it is claimed that the discovery of incommensurability did not plunge mathematics into a crisis and that far from being paralyzed mathematicians were spurred on to do more mathematics, it remains the case that there was an incoherence in the foundation of the subject. For, what is a ratio, defined as 'a sort of relation in respect of magnitudes between two magnitudes of the same kind' (*Elements*, V, definition 3), if the magnitudes are literally incommensurable? How can $\sqrt{2}$ be compared with 1? Plainly not at all if by a ratio one means a comparison between numbers. On the other hand, similarity is based on the ability to compare magnitudes. The incoherence will be resolved once a way is given to compare ratios, so that statements of the form $a:b = c:d$ make sense. Notice that each ratio can be understood on its own. The quantities a and b might both be lengths or areas or angles (i.e. of the same kind). Each has a size, and because they are 'of the same kind' they can be compared in respect of size, for example by making them overlap. The

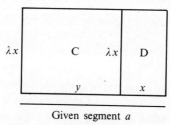

Given segment a

Fig. 2.2.

problem is to compare the resulting ratio $a:b$, not itself considered to be a number, with another ratio $c:d$. This would be no problem if all lengths were commensurable, but it becomes a problem once incommensurability has to be reckoned with.

Eudoxus' achievement was to give a completely successful set of definitions for a calculus of ratios. The definitions, as given in the *Elements*, Book V, are as follows:

Definition of ratio (*Elements* V, definition 3):

A ratio is a sort of relation in respect of size between two magnitudes of the same kind.

Ratio condition (definition 4):

Magnitudes are said to have a ratio to one another when they are capable on being multiplied of exceeding one another.

Equality of ratio (definition 5):

Magnitudes are said to be in the same ratio, the first to the second as the third to the fourth, when if any equimultiples whatever are taken of the first and the third and any equimultiples whatever of the second and fourth the former equimultiples alike exceed, alike are equal to or alike fall short of the latter equimultiples respectively taken in the corresponding order.

Comparison of ratio (definition 7):

When of the equimultiples the multiple of the first magnitude exceeds the multiple of the second but the multiple of the third does not exceed that of the fourth then the first is said to have a greater ratio to the second than the third to the fourth.

In more modern dress, the definition of equality of ratio says that if $a, b, c,$ and d are four magnitudes such that for any positive multiples m and n, which are whole numbers, $a:b = c:d$ if

$$ma > nb \quad \text{and} \quad mc > nd$$

or

$$ma = nb \quad \text{and} \quad mc = nd$$

or

$$ma < nb \quad \text{and} \quad mc < nd.$$

The definition of comparison of ratio says that $a:b > c:d$ if numbers m and n can be found such that

$$ma > nb \quad \text{but} \quad mc < nd.$$

In the case of the side and diagonal of a square we can form the ratio of the diameter d to the side s, written $d:s$, and consider how it compares with other ratios. A good approximation to $\sqrt{2}$ is $7:5$. We know that $d:s > 7:5$, but how would Eudoxus show it?

For a start, $d:s$ does not equal $7:5$, since, if it did, by the definition of equality, if $d:s = 7:5$ then whenever $m.d > n.s, m.7 > n.5$. However, taking $m = 29$ and $n = 41$ we obtain $29d > 41s$ but $29.7 < 41.5$. Accordingly, the definition of comparison of ratio allows Eudoxus to infer that $d:s$ exceeds $7:5$.

A second method to compare $d:s$ and $7:5$ is as follows. Lay d and s along a line and roll them over and over like rulers. The ratios $d:s$ and $7:5$ generate a pattern of lefts and rights which can be read off down the lines. The $d:s$ pattern is

RLRLRRLRLRRLRL...

the $7:5$ pattern is

RLRLRRLRLR B RLR...

and the $21:15$ would be the same as the $7:5$ only larger. Since the patterns for $d:s$ and $7:5$ differ we infer that $d:s$ does not equal $7:5$ and a more detailed look would show that $d:s$ exceeds $7:5$. This method has similarities with the so-called Euclidean algorithm discussed in the exercises.

The Eudoxan method allows us to take a ratio of two magnitudes and to compare it with any other. It does not explicitly resolve the existence question by exhibiting all possible magnitudes, but it does enable us to calculate with any segments we can construct by expressing them as ratios which are possibly equal to known ones or possibly unequal but in that case still comparable.[4] Books V and VI establish in detail that the method rescues similarity from its lack of definition.

Appendix

Prominent amongst the reasons for speaking of a Greek *algebra* at all is their study of equations (it would be quite possible to describe all of Euclid as geometry and arithmetic, but that is perhaps a mere terminological quibble). This study derives from the work of Babylonian writers, who, unlike their Egyptian contemporaries, were adept at solving both simultaneous equations in two unknowns and quadratics. Typically, these were expressed in pictorial language as area plus side equals a number.

The favourite form of the problem for puzzle setters was as two simultaneous equations connecting the product and sum (or difference) of two unknowns. Thus we find problems involving a number and its reciprocal (*igi, igibum*) $x.x^{-1} = 1, x+x^{-1} = b$ which is a special case of $x^2 + 1 = bx$. We also find quite generally problems like

$$xy = a$$
$$x+y = b$$

---

[4] In this sense it differs from Dedekind's definition of irrationals in *Continuity and irrational numbers* (1872) which was constructive (see Weyl 1963, p. 39).

stated, of course, in words, and equivalent to $x^2 + a = bx$, or

$$xy = a$$
$$x - y = b$$

equivalent to $x^2 = bx + a$. The quadratics are most commonly stated in this form, in which case they are solved by passing to the genuinely simultaneous linear equations

$x + y = b$ and $x - y = \sqrt{(b^2 - 4a^2)}$ in the first example
$x - y = b$ and $x + y = \sqrt{(b^2 + 4a^2)}$ in the second example.

Book II of the *Elements* makes use of a technique utterly characteristic of Greek mathematics called the application of areas. Proclus (Morrow edn, p. 332ff.) quoting Eudemus, attributes it to the Pythagoreans.

As Neugebauer has argued (1969, p. 149), the application of areas seems to have a Babylonian origin, from which it springs much changed. It is required to construct upon a given segment PQ a rectangle PQRS which is divided into two rectangles, one of given area C and the other of given shape D. If it can be done then the constructed rectangle is said to have been applied to the segment and to be deficient by D. Alternatively, the large rectangle is required to overlap the given segment in a given shape, in which case the application is said to be in excess by D. Algebraically, the two problems are these, taking the deficiency first.

The given segment has length a, say. The given shape is a rectangle of side λ by 1, say, so any copy of it is λx by x, with x along the base. If we let the given area C stand on the base of y units we have

$$\lambda xy = C$$

by looking at areas, and

$$x + y = a$$

looking along the base. Accordingly the problem reduces to finding two lengths x and y with a given sum a and a given product C/λ.

In the case of application in excess we similarly seek x and y such that $\lambda xy = C$ and $x - y = a$. Then

$$\lambda xy = C \quad \text{and} \quad x - y = a.$$

The two pairs are evidently the Babylonian form for quadratics discussed earlier. It is therefore plausible that they were reformulated in the language of geometry, resulting in the technique of application of areas.

One advantage of the appeal to geometry can now be mentioned to illustrate the gain in explanatory power. Evidently there are no numbers x and y whose sum is 10 and product 40, and the Babylonian scribes seem to have avoided discussing such questions. However, we can now see why there are no such numbers. In the language of application of areas we have to put a rectangle of area 40 on a segment of length 10 leaving a square behind.

The area xy of the large rectangle varies with x (and therefore with y) but is greatest when the rectangle is a square. In that case $x = y = a/2$, and the area is $a^2/4$. Therefore we can solve the problem provided $a^2/4$ exceeds the specified area C, i.e. $a^2/4 > C$. In our example $100/4 = 25$ is not greater than 40, so no numbers can be found. If such a problem can be solved there are then two values for x and y—can you exhibit them in a figure?

The discussion of the feasibility of finding a solution is indeed to be found in Euclid (Book VI, Prop. 27) immediately preceding the solution of the quadratics themselves (VI 28 and 29).

Exercises

2.1 It is plainly a question of historical interpretation as to whether the application of areas is regarded as 'real' algebra, i.e. the solution of equations, or whether it is inextricably geometry. Does it give insight to explain it as being about quadratic equations, or is this a misleading modernism reflecting only our greater facility with algebra and the algebraic formulation of much of (modern) mathematics? Geometrical algebra is geometrical in its methods but algebraic in purpose—is this a useful distinction to make? Can it indeed be made?

It is easy to see by drawing a large isosceles right-angled triangle that the ratio $\sqrt{2}:1$ is nearly 3/2 or 7/5 or 17/12, or certain other ratios. These ratios form a sequence called the sequence of side and diameter numbers, and the simple rule which determines them seems to have been discovered by the Pythagoreans, according to Proclus (see Heath 1956, vol. I, p. 400). Algebraically, the process is as follows.

We call the side numbers $s_1, s_2, s_3, \ldots$, the diameter numbers $d_1, d_2, d_3, \ldots$, and take for starting values $s_1 = 1$, $d_1 = 1$. We now iterate according to the rule

$$s_2 = s_1 + d_1 = 2$$
$$d_2 = 2s_1 + d_1 = 3$$
$$s_3 = s_2 + d_2 = 5$$
$$d_3 = 2s_3 + d_3 = 7$$

and in general

$$\left.\begin{array}{l} s_{n+1} = s_n + d_n \\ d_{n+1} = 2s_n + d_n \end{array}\right\}$$

(a) Each new s is the sum of the old s and d; each new d is the sum of twice the old s and then the old d.

Accordingly, $s_4 = 12$, $d_4 = 17$, and so on. To see that d_n/s_n steadily approximates to $\sqrt{2}$, we argue that $2s_n^2 - d_n^2$ is alternately $+1$ and -1, i.e. if

$$2s_n^2 - d_n^2 = \pm 1$$

then
$$2s_{n+1}^2 - d_{n+1}^2 = \pm 1.$$
We can check this as follows:
$$\begin{aligned}
2s_{n+1}^2 - d_{n+1}^2 &= 2(s_n + d_n)^2 - (2s_n + d_n)^2 \\
&= 2s_n^2 + 4s_n d_n + 2d_n^2 - 4s_n^2 - 4s_n d_n - d_n^2 \\
&= -2s_n^2 + d_n^2 \\
&= -(2s_n^2 - d_n^2)
\end{aligned}$$
and of course
$$2s_1^2 - d_1^2 = 2.1 - 1 = 1.$$
So we have $2s_n^2 - d_n^2 = \pm 1$ and therefore
$$2 = \frac{d_n^2}{s_n^2} \pm \frac{1}{s_n^2}.$$

Now s_n increases indefinitely as we increase n, being the sum of increasingly many terms each greater than unity, and so $1/s_n^2$ becomes smaller and smaller. In this way, d_n^2/s_n^2 increasingly approximates 2, and d_n/s_n likewise approximates $\sqrt{2}$. It seems certain that this method of approximating $\sqrt{2}$ yielded the astonishingly good classical estimates of $\sqrt{2}$, although the algebraic argument that $2s_{n+1}^2 - d_{n+1}^2 = -(2s_n^2 - d_n^2)$ would have been done geometrically. The geometry is, however, simple; we read in Euclid, Book II, Prop. 10 that

If a length AB be extended arbitrarily to D and AB also bisected at C then
$$AD^2 + DB^2 = 2AC^2 + 2CD.$$
Let AC $= x$, BD $= y$, then AD $= 2x + y$, etc. and in symbols
$$(2x+y)^2 + y^2 = 2x^2 + 2(x+y)^2.$$
By rearranging this as
$$2x^2 - y^2 = 2(x+y)^2 - (2x+y)^2$$
and taking $x = s_n$, $y = d_n$
$$x + y = s_{n+1}, \qquad 2x + y = d_{n+1}$$
we see that this is the result we want.

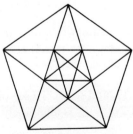

Fig. 2.3.

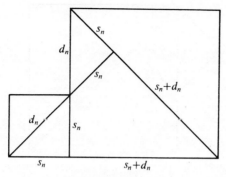

Fig. 2.4.

2.2 By taking initial values for s_1 and d_1 of, say, 1 and 2 respectively, show that it does not matter what initial values for s and d are taken, at least over a certain range.

2.3 By replacing the 2 in the iterative formula by another number show that the same method will calculate any square root. Use $s_{n+1} = s_n + d_n$, $d_{n+1} = 5s_n + d_n$ to calculate approximations to $\sqrt{5}$.

2.4 Use the same method to calculate approximations to the golden section by setting $s_{n+1} = s_n + d_n$ and $d_{n+1} = s_n$. How does this relate the inscribed pentagons.

A good discussion of Greek use of this material is to be found in Heath (1956, especially Vol. I, pp. 395–401).

2.5 Side and diameter numbers can frequently be displayed in diagrams. The rule $s_{n+1} = s_n + d_n$, $d_{n+1} = 2s_n + d_n$ can be displayed as follows. Construct a square of side s_n and diameter d_n. Extend one (vertical) side by d_n and construct a new square on the side (of length $s_n + d_n = s_{n+1}$). Prove that the diameter of the new square is $2s_n + d_n = d_{n+1}$. (The figure is full of isosceles right-angled triangles.)

If the formulae are run backwards, so that steadily smaller s and d are produced according to the rule $s_n = d_{n+1} - s_{n+1}$, $d_n = 2s_{n+1} - d_{n+1}$, or, if you prefer, $s'_{n+1} = d'_n - s'_n$, $d'_{n+1} = 2s'_n - d'_n$ for a new sequence s', d', the sequence of squares shown in Fig. 2.5 is produced. Compare this process with the Euclidean Algorithm (*Elements* X, 3) for finding the greatest common measure of two commensurable magnitudes. Let the magnitudes be a and b, with $a < b$. Then

$$b = n_1 a + r_1$$

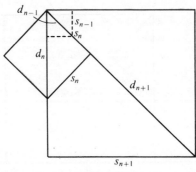
Fig. 2.5.

where $0 \leq r_1 < a$, for some number n_1. Similarly

$$a = n_2 r_1 + r_2$$

where $0 \leq r_2 < r_1$ and so on until eventually

$$r_k = n_{k+1} r_{k+1}$$

where r_{k+1} is the greatest common measure. For example

$b = 17, \quad a = 3$
$17 = 5 \times 3 + 2$
$3 = 1 \times 2 + 1$
$2 = 2 \times 1$ 1 is the greatest common measure

and

$20 = 1 \times 14 + 6$
$14 = 2 \times 6 + 2$
$6 = 3 \times 2$ 2 is the greatest common measure.

2.6 To modern mathematicians, but not the ancients, it was an acceptable proof of incommensurability that the Euclidean algorithm never terminated. Show that the process for producing s's and d's above is a Euclidean algorithm which never terminates. Deduce that $\sqrt{2}$ is irrational. Deduce the same for $\sqrt{5}$ by means of Fig. 2.3, taking s as a side of the regular pentagon and d as a diagonal.

Magnitudes

2.7 One candidate for magnitudes of different kinds in our sense is angle, for the Greeks admitted angles between curved lines *not* in the modern sense of angles between their tangents. The curious proposition in Euclid III, 16 that the angle between the circle and its tangent is less than any rectilineal angle is an example of this. By consulting, say, Heath (1956) try to decide for yourself if indeed such angles can be considered as magnitudes. Magnitudes can be added. How would you define the addition of angles between curved lines? (In this connection see the end of the book by Klein (1939).)

3 Euclidean geometry and the parallel postulate

Preamble: motion and space

Geometry is not, of course, a branch of arithmetic. It rapidly ceased to be subordinate to arithmetical questions and came to rank alongside it as an inquiry of equal rigour and value, its subject being figures in space and the properties of those figures. At first the ambient space was taken as a plane, but later three-dimensional figures were considered. The Greek geometrical proofs worked because of assumptions made about the underlying space, which are reflected in the ideas of congruence, similarity, and parallelism, and in the ability to make geometrical constructions.

The basic notion is that of congruence. Two figures are *congruent* if they can be made to coincide exactly with one another, in which case they are sometimes said to be equal in all respects. The method of establishing this coincidence is, at least in principle, the movement of one figure until it is placed exactly upon the other, alternatively, it may be established that the data for constructing the two figures are the same, i.e. the instruction kit, so to speak, makes only one figure. In the example of a congruence theorem attributed to Thales it was claimed that given a segment and two angles standing upon it, a unique triangle can be constructed for any given position of the segment and that the construction carried out upon equal segments in different positions yields congruent triangles. It might be that two triangles constructed in this fashion are mirror images of each other, in which case motion must be understood to include a reflection, but with this understanding one can state that Greek geometers assumed that the (idealized) space in which their (idealized) figures existed was homogeneous for a given construction can be performed anywhere in it with the same results each time. The kinds of motions which were admitted were those which respected the rigidity of a triangular frame: rotations, translations, and reflections. A triangle neither warps nor contracts as it is moved about, for if it did the whole notion of congruence would collapse. Of these motions, consider only translations, i.e. motions in a given direction by a given amount. A translation of AB by t units north east moves A and B equally by that amount, to A' and B' say. It leaves the segment AB unchanged, so that A'B' = AB, and of course AA' is parallel and equal to BB'. Therefore opposite sides of a parallelogram are equal and parallels are everywhere the same distance apart. Or are they? One great achievement of Greek geometry was to build up a system of geometry sufficiently subtle to enable such problems to be posed and answered without fear of begging the question. The formulation of the concepts largely suppressed questions of rigid-body motion in a homogenous medium in

30 *Euclidean geometry and the parallel postulate*

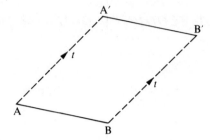

Fig. 3.1.

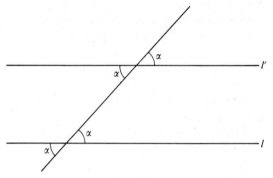

Fig. 3.2.

favour of the concept of congruence and the ability to make certain constructions arbitrarily in space. Rather than discuss translations directly, the Greeks preferred to work with parallel lines, which, one might say, are the tracks along which a translation is performed.

As a transformation of a figure which is not a congruence consider a similarity, the production of a scale copy. Figures which are similar have, as we have seen, corresponding pairs of angles equal and corresponding ratios of sides equal. It is again assumed that the ambient space permits us to construct scale copes of given figures.

Let us now consider the most famous formulation of geometry ever given.

Euclid's Geometry

Euclid's *Elements* opens with 23 definitions, in which most of the basic terms are defined, five postulates, and five common notions. The postulates permit certain geometrical constructions to be made: to join up two points with a line, to draw a circle of any radius with centre any point, and so on. The common notions are permissible deductions or rules of inference applicable outside of mathematics: things which are equal to the same thing are also equal to one another, and so on.

Not all of the definitions are of equal worth. The opening ones attempt to define point, line, straight line, surface, and plane, which are perhaps better left undefined. Aristotle argued that every subject must start from indemonstrable principles, and these terms could be taken as primitive; even so he did define a point (in the *Metaphysics*, 1916b) as that which is indivisible and has position, while in Euclid it is that which has no part. On either definition it is hard to see how a collection of points can make up a line (a 'breadthless length' according to Euclid) and indeed the Greeks never regarded a line or curve as the successive positions of a moving point. Geometric figures were to be defined independently of ideas of motion. Even as simple a curve as a circle is defined as

a plane figure contained by one line such that all the straight lines falling upon it from one point lying among those lying within the figure are equal to one another.

The special point is the centre of the circle. We also know from Aristotle, who was fond of mathematical allusions, that earlier compilers of elementary geometries (e.g. Theudius) built them up from the same basic figures: points, lines, and circles all in one plane. Once these terms are defined we can reasonably define others: triangles, quadrilaterals, and so on, and, as special cases, squares, rectangles, and the like. We also know that one definition and postulate attracted more attention and controversy than all the others: postulate 5 on parallels.

As we have already said, parallels are straight lines in the plane which never meet even if produced indefinitely. The natural question arises as to whether there are any parallel lines, for not even a mathematician can create merely by defining. However, without parallels it is hard to do very much geometry at all, because parallels are needed to transport equal angles about. For instance, the angles in Fig. 3.2 are equal because l and l' are parallel.

Table 3.1. Euclid's Elements

Postulates
Let the following be postulated.
(1) To draw a straight line from any point to any point.
(2) To produce a finite straight line continuously in a straight line.
(3) To describe a circle with any centre and distance.
(4) That all right angles are equal to one another.
(5) That, if a straight line falling on two straight lines makes the interior angles on the same side less than two right angles, the two straight lines, if produced indefinitely, meet on that side on which are the angles less than the two right angles.

Common notions
(1) Things which are equal to the same thing are also equal to one another.
(2) If equals be added to equals, the wholes are equal.
(3) If equals be subtracted from equals, the remainders are equal.
(4) Things which coincide with one another are equal to one another.
(5) The whole is greater than the part.

Taken from Heath (1956, Vol. I, pp. 154, 155).

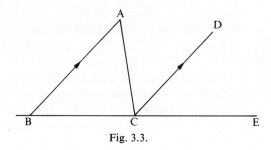

Fig. 3.3.

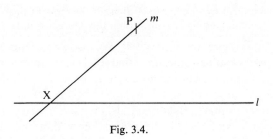

Fig. 3.4.

It follows that the angles in any triangle add up to 180°; here is Euclid's proof (*Elements*, I, Prop. 32). ABC is any triangle; CD is drawn parallel to BA and BC is extended to E. We have $B\hat{A}C = A\hat{C}D = \hat{A}$, say, and $A\hat{B}C = D\hat{C}E = \hat{B}$, say, because BA is parallel to CD. Looking at the angles at C we have, calling $B\hat{A}C = \hat{C}$, $\hat{A}+\hat{B}+\hat{C} = 180°$, and so the angles in the triangle, also $\hat{A}+\hat{B}+\hat{C}$, add up to 180°.

Since geometrical figures are studied in terms of distance and angle it is clear that parallels are useful tools to have around; the arguments just given occur immediately after parallels are introduced into the *Elements*, Book I. Therefore the question arises as to the existence of parallels. The usual demonstration is the 'seesaw proof'. Take a line *l* and a point P not on it. Draw a line *m* through P, which can be thought of as a seesaw pivoted at P above the ground *l*; admittedly it is a seesaw of indefinite extent. If the line *m* already never meets *l* then it is parallel to *l* and the existence question has been resolved. Therefore suppose that it is not parallel, but meets *l* as in Fig. 3.4 at X. Tilt *m* until it meets *l* on the other side of P, at Y say, and consider what happened to X.

We can be sure that *m* does not meet *l* to the left and right at the same time, since then *m* and *l* together would have enclosed an area, and so *m* and *l* would be two different straight lines between two points, X and Y. This involves a contradiction with the premises of geometry, and so can be rejected. Is it still possible that there could be no parallels? As we tilt *m* it seems clear that X moves away leftwards, until *m* and *l* part company to the left, and it is surely impossible that there could be a position of *m* in which *m* met *l* at X (to the left), but topple it ever so slightly and it meets *l* at Y (to the right). More plausibly, it parts company with *l* leftwards before it joins up rightwards. Take the symmetrical position in which all angles are right angles. If *m* meets *l* at all, then does it do so to the left or at the right? Surely by symmetry it must do so either at both or at neither, and the possibility of both X and Y has just been refuted.

Therefore *m* can be drawn parallel to *l*. However, the seesaw argument now suggests an embarrassment of riches: why should *m* not part company with *l* to the left long before rejoining it at the other side, providing us with a whole spread of positions in which it never meets *l*? We could imagine that *m* draws steadily nearer *l* without ever meeting it, and the Greeks themselves referred to curves which did that. Thus the arc of a rectangular hyperbola draws nearer to its axes without ever meeting them (Proclus, Morrow edn, p. 151). Generally, two curves which arbitrarily approach one another without meeting are said to be asymptotic, or to approach one another asymptotically, so, rephrasing our point, it is not obvious that straight lines cannot be asymptotic but rather it is something that ought to be proved.

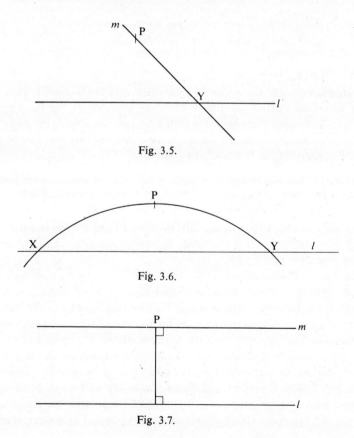

Fig. 3.5.

Fig. 3.6.

Fig. 3.7.

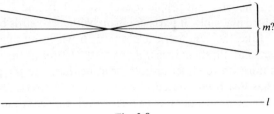

Fig. 3.8.

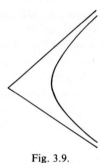

Fig. 3.9.

The parallel postulate

Strangely enough, the proof sought after was never found. However, if a parallel m to l can be wobbled, then, as a glance at the theorems already quoted shows, parallels are no use for transporting angles. Parallels must not only exist, but they must be unique if they are to be any use. Euclid got out of this by means of his famous fifth postulate:

If a straight line falling on two straight lines make the interior angles on the same side less than two right angles the two straight lines, if produced indefinitely, meet on that side on which are the angles less than the two right angles.

Therefore in Fig. 3.10, where n falls on m and l and $\alpha + \beta$ is less than two right angles, m meets l, when extended, somewhere over to the right and on the same side of n as α and β.

In terms of the seesaw it is no loss to replace n by n', meeting l in a right angle. It is clear from looking at P that $\alpha - \alpha' = R - \beta$ and so $\alpha + \beta = \alpha' + R$.

What Euclid now postulated was that we can infer from the inclination of m to n' that m will meet l as it seems to, to the right, the side on which it makes the smaller angle with the vertical. However, he had to assume this, the act of making it a postulate having this arbitrariness about it. If we want to try and prove that m still meets l when it is symmetrically poised, then we can, but Euclid simply excluded this possibility by fiat. In so doing he concurred with most geometers before him and since. None the less, the problem of parallels puzzled Greek geometers a great deal. For a start, the postulate

Euclidean geometry and the parallel postulate   35

is non-intuitive and asserts things about lines meeting lines indefinitely far in the distance. It may seem odd that m and l could part company for a whole range of positions of m, but this is geometry, not mere reality, and we need proofs of things. Having to settle for a Gordian solution to the problem is rather unsatisfactory and possibly even misleading. Perhaps the postulate could even give trouble.

We must distinguish here between the bizarre results and contradictions which follow from making an assumption. If it is assumed that there is a number x such that simultaneously $x+1 = 2$ and $x+1 = 3$ then we have a contradiction: $2 = 3$. Since $2 = 3$ is false it follows that there can be no x with the desired properties. If, however, we assumed there is a number x such that $x^2 - 10x + 40 = 0$ we might or might not be in trouble. We would nowadays obtain as solutions $x = 5 + \sqrt{(-15)}$ and $x = 5 - \sqrt{(-15)}$, and would not worry about the $\sqrt{(-15)}$. However, there was a time when the square roots of negative numbers were considered self-evident nonsense and even negative numbers were called fictitious. By about the sixteenth century numbers like $\sqrt{(-15)}$ were admitted, as 'imaginary', and in the prevailing

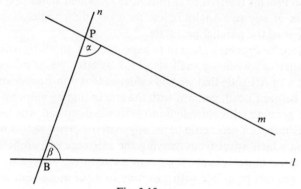

Fig. 3.10.

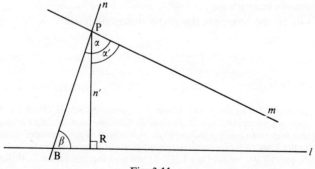

Fig. 3.11.

36 Euclidean geometry and the parallel postulate

theory were considered bizarre but not self-contradictory, the number system and algebra being enlarged to handle them in a way that could not have happened with a number x with $2 = x+1 = 3$. The algebra of imaginary quantities was merely bizarre; in the words of one of its architects 'as subtle as it is useless'.[1] In the history of research into parallels we shall find the same range of attitudes and responses.

The parallel postulate was notorious even before Euclid; it grated that such an assumption had to be made. The prevailing belief was that it surely followed from the straightness of the lines that lines like those in Fig. 3.10 *had* to meet, and so the postulate would turn out to be unnecessary. The debate ran in three channels: attempts to derive the postulate from the rest of elementary geometry; attempts to reformulate the postulate or the definition of parallels into something less objectionable; descriptions of what geometry would entail if the postulate was in some way denied. Proclus (Morrow edn, p. 150) wrote of it:

This ought to be struck from the postulates altogether. For it is a theorem ... and requires for its demonstration a number of definitions as well as theorems. And the converse of it is proved by Euclid himself as a theorem.

The converse Proclus referred to is Euclid, I, 17, which states that in any triangle the sum of any two angles is less than two right angles, a result which is independent of the parallel postulate.

We shall look at Proclus's attempt to expel the hypothesis by making it into a theorem later on, when we shall also look at attempts to reformulate the postulate. It is in Aristotle that we find evidence that non-Euclidean geometry was studied before Euclid, perhaps with the aim of finding amongst its bizarre properties a genuine self-contradiction. Aristotle discussed whether thinking that parallels meet is a geometric or an ungeometric error, that is whether the contradiction which arises from denying the existence of parallels is strictly mathematical or more broadly logical in its nature.[2,3] He also connected denying the parallel postulate with asserting that the angle sum of a triangle is greater than 180°. Here he was evidently referring to a considerable body of knowledge about the problem which is now lost. We shall see below how these connections can be made, so let us only note here that they were made as early as the fourth century B.C.

We turn now to the attempts upon the postulate itself.

---

[1] Cardano, *Ars Magna* (*The great art*), Chap. 37, quoted by Struik (1969, p. 69).

[2] We might discuss a misapplication of statistics in the same way: is a man wrong to apply to statistics in this way or to analyse the problem in this way? Proponents of racial theories of intelligence make both mistakes, and so, unhappily, do their opponents.

[3] Heath (1949, pp. 57, 41, = An Post I, 12, 5; see also the articles by Toth listed in the Bibliography.

Immediate consequences of the postulate

In Euclid, parallels are first used in Book I, Prop. 27, when he proves that if a straight line crosses two others in such a way that the two alternate angles are equal, then the two lines are parallel (see Fig. 3.2).

This is another symmetrical position for the seesaw, and we can take it as a method for constructing parallels. Given a line l draw n, any n, making an angle with l, and through a point of n draw the line m which copies that angle.

He then proved the usual properties of angles and parallels. A typical result is the 'transitive' property of parallelism: if a is parallel to b and b is parallel to c then a is parallel to c. The foundations of the study of parallels are completed by showing in Prop. 31 that through a given point P not on a line l a line m can be drawn parallel to l. In our argument above, take care to draw n through P. Now m must be the only parallel to l through P, since, if m' is another, then m is parallel to l, so m' is parallel to l and m is parallel to m'. However, m and m' meet at P, which is absurd unless m and m' are the same line. Therefore we have not only constructed a parallel to l through P, but indeed the unique parallel to l through P. We can summarize this whole discussion of parallels by saying that, in Euclid, given a line and a point not on it there is a unique parallel to the line through that point. In this form the postulate is sometimes known as *Playfair's axiom*, after John Playfair who brought out successful editions of Euclid in the years following 1795. It can, however, be found in Proclus (Morrow edn, p. 295) in his comments on Prop. 31. None the less it is the clearest summary of the postulate to modern eyes, and we shall follow most recent commentators in regarding it as the best statement of Euclid's position. Its great attraction is that it can readily be reformulated to suggest non-Euclidean geometries by denying either the existence or uniqueness of parallels.

In classical times the emphasis in research was on proving the uniqueness of parallels, existence being taken for granted. Some attempted to show that the postulate could be proved as a theorem: others attempted to express it in less objectionable ways, usually by appealing to the concept of equidistance. Think of Euclidean geometry without parallels as a body of evidence—admittedly slight—and the parallel postulate as a new piece upon which we have to check. Is it consistent with what we know already? It might turn out that what we know already compels us to deduce the validity of the new piece anyway. It might turn out that denying the postulate results in contradictions with what is already known, in which case we again have to accept it, or it might be too intractable, in which case we look for a new piece of evidence to clinch the case. The aim in each case is to graft onto geometry without parallels (an unexceptionable but dull subject) something unexceptionable but powerful enough to make the subject interesting.

At least informally parallel lines are always the same distance apart. However, as we have seen with curves, a steady drawing together need not imply an eventual meeting. There is an ingenious classical argument designed

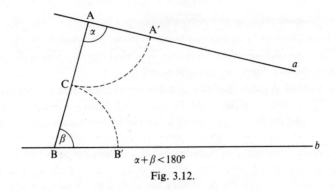

Fig. 3.12.

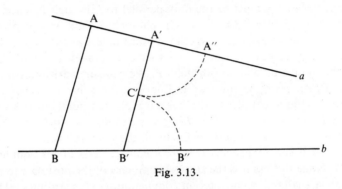

Fig. 3.13.

to show that such lines as *a* and *b* in Fig. 3.2 do not meet. It is stated by Proclus (Morrow edn, p. 289), who does not succeed in refuting it although he commented pertinently. The argument, as has often been remarked, recalls the paradoxes of Zeno.

Let *a* and *b* be two lines which cross a third, *c*, at A and B respectively and mark off interior angles α and β such that $\alpha + \beta < 180°$. The argument intended to show that *a* and *b* do not meet says bisect AB at C, and locate A' on *a* and B' on *b* such that

$$AA' = AC = CB = BB'$$

in length. *a* and *b* cannot meet between A and A' or B and B', since if they did they would create a triangle (with sides *a*, *b*, and *c*) and the side AB would exceed the other two put together, which is plainly impossible, as is proved in Euclid Book I, Prop. 20. Therefore *a* and *b* do not meet between A and A'. Draw A'B' and cut it at its midpoint C'. Locate A'' on *a* and B'' on *b*, such that

$$A'A'' = A'C' = C'B' = B'B''.$$

Exactly as before, *a* and *b* cannot meet between A' and A'', B' and B''. Therefore continue in this fashion with C'', A'', B'', and it can be kept up indefinitely, producing an infinite sequence of points, A's and B's. In conclusion, says our anonymous theorist, a and b can never meet, because there is an infinite sequence of segments between A and any conceivable meeting point.

The fallacy is of course that the infinite sequence produced here converges, but that is not really the point of interest. More importantly, at least the drawing together of a and b can be seen to require a decent proof. What Proclus (Morrow edn, p. 290) says about it is even more interesting. He took $\alpha + \beta < 180°$ and supposed that a and b do not meet; he then joined A to an arbitrary point B' on b with $B\hat{A}B' = \gamma$, say. Evidently γ is smaller than α, so $\gamma + \beta < 180°$. Therefore it is false, he said, that whatever angles two lines make with a third, provided that their sum is less than 180°, the lines cannot meet since AB' and b do. Therefore it can only be the case that there is some value for the sum $\alpha + \beta$ below which the lines do meet and above which they do not. This would provide a with a range of positions in which it never meets b, a wobble to the seesaw. Proclus did not go further than this, but we shall later when we discuss the work of Lobachevskii and others. (A question to be getting on with: does the size of the wobble at A depend on the height AB?)

Distance between lines

Proclus (Morrow edn, p. 291) attempted to show directly that the postulate is a theorem, basing his argument on an explicitly Aristotelian appeal to the finiteness of the universe. His argument is in three parts:

(1) In any angle, he said, the interval between the lines forming the angle 'will exceed any finite magnitude' (this corresponds to the assumption of Aristotle, in *De Caelo* 271b28ff, that the radius of an orbit can only be finite). The lines a and b in Fig. 3.15 make an angle at P, and the distances $d_1, d_2, d_3, \ldots$ between a and b increase without limit as you move away from P.

(2) A result about parallels:[4] let m and l be parallel and n cross m at P. Then,

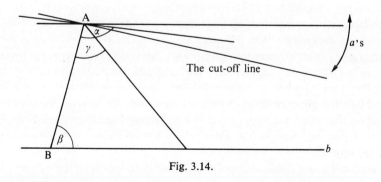

Fig. 3.14.

---

[4] For Proclus, *Elements* I, 27 guarantees the existence of parallels; it is uniqueness that must be proved.

40 *Euclidean geometry and the parallel postulate*

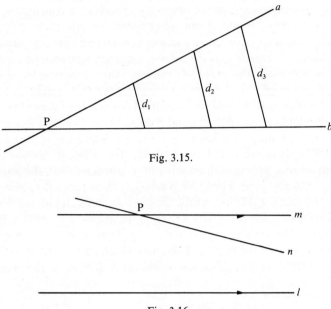

Fig. 3.15.

Fig. 3.16.

claimed Proclus, n must also meet l. Indeed, the distance from n to m increases without limit as we move to the right and so must eventually exceed the distance from m to l, but if n moves beyond l it must at some stage have crossed it and so n meets l.

(3) Now we deduce the postulate in its original form. Let k cross m and l, and let the angles α and β be such that $\alpha + \beta < 180°$. We must show that m meets l. Through P draw m' parallel to l; m' makes an angle with k of γ, say, and we know $\beta + \gamma = 180°$, so m crosses inside m' at P, and so, from (2), it must meet l and the result is proved.

Although Proclus had explicitly admitted as an assumption that the distance between two intersecting lines increases indefinitely, his proof is not without its deceptions. The difficulty occurs in (2). It is clear that n moves steadily away from m (clear by assumption), but it is not clear that the distance from m to l does not change as well. We have met this possibility earlier, when we called it the equidistance of parallel lines, and it should at least be explicitly stated that the distance between parallel lines does not increase without limit. Based on these two assumptions, that intersecting lines diverge without limit and that parallel lines do not, Proclus has indeed shown that parallels exist and are unique.

Significantly, no classical proof of the postulate was accepted as conclusive by the ancients. At best, Proclus gave a reformulation of it, and it is in terms of distance that most people after him sought to tackle the problem. Simplicius, a sixth century A.D. commentator on Aristotle, also wrote a commentary

Euclidean geometry and the parallel postulate 41

on Euclid's first book in which he gives a demonstration due to his friend Aganis.

The equidistant curve

Aganis assumed that the curve everywhere equidistant from a straight line is itself straight. This is not obvious, as can be seen by considering a globe; although there are no straight lines on a sphere, part of the equator is the curve of shortest distance between any two of its points, but no other line of latitude is the curve of shortest distance between any of its points. Let us *assume*, however, that in plane geometry the curve equidistant from a line is straight. Then it is evidently parallel to that line, since they do not meet. It remains to check that parallels are unique, i.e. that the only non-intersecting line is the equidistant one. In the familiar picture, m and l meet, where n has been drawn perpendicular to l. Take a point M on m and draw the perpendicular to n, MN say. If AB is bisected at B′, and AB′ bisected at B″ and so on we obtain a sequence of points B′, B″, ... which draw nearer and nearer to A. Eventually one of them, say $\tilde{B}$, must be nearer to A than N is. Draw $\tilde{B}\tilde{M}$ perpendicular to n, meeting m at $\tilde{M}$. Now on m draw AC, the same multiple of A$\tilde{M}$ as A$\tilde{B}$ is of AB. Then m and l meet at C, as was required. Effectively this is a similar triangle argument; the triangle on A$\tilde{M}$ is expanded until A$\tilde{M}$ equals AB.

Aganis's argument does indeed follow from his premises, but it is based upon the existence of equidistant lines which is a strong assumption to make. To be precise, if we assume with Aganis that the curve equidistant from a straight line is itself straight then the two lines are parallel, and if P is a point not on a line l there is exactly one parallel to l through P. It is the line through P which is everywhere the distance from l that P is, which we might call the railway line.

Conversely, if we define parallels as Euclid does and accept his fifth postulate, then of course, parallel lines are like railway lines, everywhere the same distance apart. (The appropriate Proposition is that opposite sides of a parallelogram are equal (Book I, 34).) Therefore we can accept either equidistance or the fifth postulate as equivalent starting points, but need not

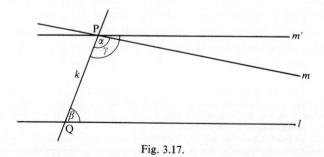

Fig. 3.17.

42   *Euclidean geometry and the parallel postulate*

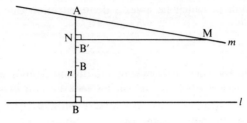

Fig. 3.18.

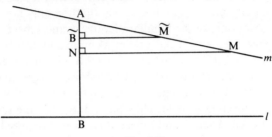

Fig. 3.19.

regard either as sufficiently elementary starting points. Characteristically, Euclid failed to draw attention to the importance of this theorem for the study of parallels.

We have Aganis's argument because the Arabian mathematician Al-Nirizi included it in a ninth century commentary on Euclid. In the thirteenth century the same criterion was again employed, this time by Nasir Eddin Al-Tusi.[5] It is due to the enthusiasm of the Arabs in collecting documents that we have as much Greek mathematics and philosophy as we do, and they themselves were also excellent creative mathematicians. When we turn to the mathematical renaissance in the West we shall see that John Wallis thought it worthwhile to have Nasir Eddin's work on parallels translated into Latin.[6]

Nasir Eddin's basic assumption is that, if a line l is perpendicular to n at A and m is oblique to n at B, then the perpendiculars from m to l are smaller than AB on the side where the smaller angle is located. Furthermore, the perpendiculars from m to l on the other side are longer than AB.

---

[5] Mohammed ibn Mohammed ibn Al-Hasan Abu Jaffa Khawaja Nasir Eddin Al-Tusi (Mohammed the son of Mohammed the son of Al Hasan, nicknamed Abu Jaffa, 'FRS', bringer of victory to the faithful, whose family name is Al-Tusi), born 1201, died 1274, assisted in the Mongol sack of Baghdad in 1258.

[6] The famous Omar Khayyam also tackled the parallel postulate on the basis of a lost premise of Aristotle which denied the possibility of asymptotic straight lines.

Euclidean geometry and the parallel postulate   43

On this assumption, what if m and l are such that two perpendiculars from m to l are the same size? Then all perpendiculars must be the same length since they cannot diminish rightwards from BA or leftwards from B'A'. Therefore the lines are equidistant, and AA'B'B is a rectangle. At this point Nasir Eddin does not follow Al-Nirizi, but does something new. By dividing AA'B'B along a diagonal he shows that in a right-angled triangle the three angles add up to 180°. From this it follows that in any triangle the angle sum is 180°; indeed by dropping the perpendicular XW from X to YZ we obtain two right triangles, XYW and XZW, and

$$\hat{X} + \hat{Y} + \hat{Z} = \hat{X}_1 + \hat{X}_2 + \hat{Y} + \hat{Z}$$
$$= \hat{X}_1 + \hat{Y} + R + \hat{X}_2 + \hat{Z} + R - 2R$$

but

$$\hat{X}_1 + \hat{Y} + R = 2R$$

and

$$\hat{X}_2 + \hat{Z} + R = 2R$$

because the two smaller triangles are right angled, and so

$$\hat{X} + \hat{Y} + \hat{Z} = 2R + 2R - 2R = 2R.$$

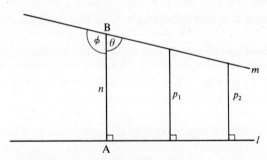

Fig. 3.20. $\phi > \theta$ implies BA $> p_1 > p_2 > \ldots$.

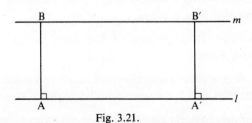

Fig. 3.21.

44 *Euclidean geometry and the parallel postulate*

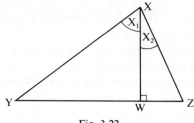

Fig. 3.22.

Finally Nasir Eddin deduced the parallel postulate from his result about the angle sum of a triangle, thus completing the circle of equivalences:

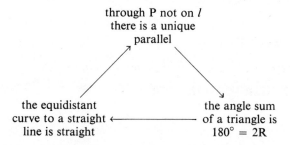

I shall omit his proof of this last step because it closely resembles that of Aganis, although it is more thorough. Note that his basic assumption concerning parallels would not seem to be as strong as that of Aganis, since on the face of it *m* could still approach *l* asymptotically. We shall see that the steady diminution in the perpendiculars for all lines *m* is ultimately the same, however. Let us also note that he foreshadowed many Western mathematicians in his attention to the angle sum of a triangle.

So we have seen that curved lines may always approach one another and yet never meet, like the hyperbola and its asymptote, so whether or not two straight lines may approach one another asymptotically becomes an interesting question. Euclid had to assume that they could not, and other classical writers could only rephrase the problem. Here are some of their versions.

If parallel lines are always a finite distance apart but intersecting lines always diverge then parallel lines are unique (Proclus).

If all triangles have an angle sum of 180° then parallels are unique, and conversely (Nasir Eddin).

Parallel lines are everywhere equidistant (Euclid) and if the equidistant curve is itself straight then it is the unique parallel at that distance (Aganis).

If we think of the paths of rays of light in empty space as being straight lines, these theorems become plausible results in optics. The doubt remains, however, because the light rays could actually bend slowly according to some rule. Alternatively, could we mathematically, and without recourse to experiment, prove that the curving of light was impossible?

It would seem characteristic of mathematics at its best that it calls forth novel investigations of its problems. Throughout the great period of its study in Greece it had this creative aspect, that it was to be elucidated by patient study. It flowered under the Moslems, and again when mathematics revived in the West. The study of Euclidean geometry only became critical late in the Western revival of mathematics, after the successful invention of the calculus for instance, but the final consequence in all of mathematics.

Appendix

Solid geometry

It is usual in mathematics to begin with a study of plane geometry, in which figures inhabit a two-dimensional space, and then consider solid or three-dimensional geometry which is much more difficult. This approach follows the historical order; however, as Plato pointed out,[7] solid geometry is the branch of mathematics most relevant to astronomy. It seems that its study went through three overlapping phases.

Firstly problems in it were reduced to problems in plane geometry by an astute choice of planes, say OAB, OAC and OBC in Fig. 3.23. This method cannot cope easily with all the problems of solid geometry, and so the next two methods were introduced.

To express the angular separation of A, B, and C as seen from O, it is convenient to regard them as centred on a sphere centred at O; this viewpoint is, of course, natural in observational astronomy, especially when distances from O are not considered especially relevant. It is first observed that the

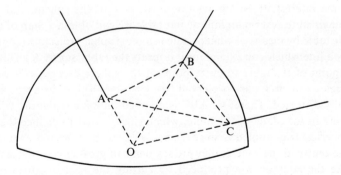

Fig. 3.23.

[7] *Republic*, VII, 529, '... the ludicrous state of research in solid geometry ...'.

46 Euclidean geometry and the parallel postulate

arc lengths CA on the sphere are directly proportional to the angles COA: $\widehat{CA} = R \cdot \widehat{COA}$, where $\widehat{CA}$ denotes arc length and R the fixed but arbitrary radius of the sphere. If $\widehat{COA} = b$, say, we shall write, in units for which $R = 1$, $\widehat{CA} = b$. Then the techniques of spherical geometry were developed, notably by Hipparchus (c. 150 B.C.) and Ptolemy (c. A.D. 125), to provide formulae relating the sides and angles of a spherical triangle. Since any discussion of Greek trigonometry would run to considerable length I shall at once state the basic formulae of spherical trigonometry in modern dress. Let the triangle ABC have angles $\hat{A}$ at A, etc., and sides $\widehat{BC} = a = \widehat{BOC}$, etc. The chief formulae are

(i) $\cos a = \cos b \cos c + \sin b \sin c \cos \hat{A}$

and

(ii) $\cos \hat{A} = -\cos \hat{B} \cos \hat{C} + \sin \hat{B} \sin \hat{C} \cos a$

which each appear in two other guises when a, b, c and $\hat{A}, \hat{B}, \hat{C}$ are replaced (in order) by b, c, a and $\hat{B}, \hat{C}, \hat{A}$ or by c, a, b and $\hat{C}, \hat{A}, \hat{B}$.

Formula (i) can be thought of as expressing $\hat{A}$ in terms of the sides a, b, c and formula (ii) likewise gives a in terms of the angles $\hat{A}, \hat{B}, \hat{C}$. There is also a formula involving sines:

$$\frac{\sin a}{\sin \hat{A}} = \frac{\sin b}{\sin \hat{B}} = \frac{\sin c}{\sin \hat{C}}.$$

As with plane triangles a spherical triangle is specified, up to position, once three pieces of data are given; indeed, a spherical triangle is known once its angles ABC are known. Accordingly these formulae can be derived from each other, and, as in plane trigonometry, it suffices to derive them in the special case of a right-angled triangle, but I shall not do this (see Chapter 10).

An intermediate stage in the development of spherical trigonometry is of some interest. If N denotes the north pole of the sphere, and the sphere is imagined to rest on an infinite flat table we can obtain a map of the sphere on the table by means of what is called stereographic projection; join N to A, say, by a line which you extend until it meets the table, say at A'. This map assigns a point of the table to every point of the sphere except N. It turns out that angles and circles are preserved (see Chapter 14) but distances are systematically distorted. Therefore ABC is mapped onto a circular arc triangle A'B'C' lying in the plane of the table, which can be used to deduce the formulae of spherical trigonometry. Since the arcs AB, etc. have been drawn with centre O, the centre of the sphere, they are segments of great circles (those circles which, like the equator, are produced by cutting the sphere with a plane passing through O). They therefore form the curves of shortest length between their end-points and are the analogues of straight lines for the geometry of the sphere.

Exercise

3.1 (To be discussed in the sequel.) In this geometry is it true that the angles of a triangle add up to $2R$? Is it true that the equidistant curve to a great circle is a great circle? Can two great circles be parallel? Can two similar but non-congruent triangles be drawn on a sphere?

Part 2

4 Saccheri and his Western predecessors

In Part II I shall carry the story of the problem of parallels on from its rediscovery in Europe to its resolution. The standard reference here is the excellent work on Bonola (1912 (reprinted 1955)), and it would be impossible to record every point at which I have consulted it. I place certain emphasis differently to him; in particular I have played down the axiomatic problems raised by the geometers because I believe them to be historically late. The interested reader is referred to the relevant sections of Bonola for a full treatment of that aspect of the subject. The book by Coolidge (1940 (reprinted 1963)) has also been helpful, and the polemical and thorough, if sometimes inaccurate, books of Morris Kline, notably his recent *Mathematical thought from ancient to modern times* (1972), should be consulted for views which I cannot share. The recent book *Euclidean and non-Euclidean geometries* by M. J. Greenberg (1974) goes into more detail on the developments in classical geometry and axiomatics than I have space for, but it does not deal with the trigonometry that played such an important role in the successful work of Bolyai and Lobachevskii.

The rediscovery of geometry in Europe in the renaissance, largely inspired by the translation of Arabic versions of Greek texts and the publication of scholarly editions in Greek and Latin of standard authors, led eventually to a new interest in the problem of parallels. A Greek edition of Proclus was brought out by Simon Grynaeus the elder in Basel in 1533 as an appendix to his edition of Euclid; a much better Latin edition by Barocius followed in 1560. Attempts to prove the postulate were made by Commandino in his 1572 edition of Euclid, Clavius in his 1574 Latin edition, Cataldi in his *Operetta delle linee rette equidistanti et non equidistanti* of 1603, also published in Latin in the same year, and Borelli (*Euclides restitutus*, 1658). Cataldi's hypothesis that straight lines which are not equidistant converge in one direction and diverge in the other is perhaps the most original, since the others merely took up the hypothesis of equidistant straight lines, but no good work was done on the problem until the mid-seventeenth century when John Wallis of Oxford took it up. He was followed by Giordano Vitale, Gerolamo Saccheri, and then by Johann Lambert, but each time the problem survived and continued to cast its shadow further over the domain of knowledge.

John Wallis (1616–1703)

In the Euclidean geometry there are two distinct ways in which two triangles may be related to each other: they may be the same shape and size (congruent) or the same shape only, when they are said to be similar. Two similar figures are copies in the sense that enlargement and reduction copy, and it is the

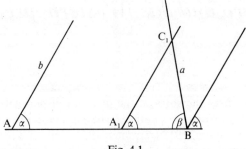

Fig. 4.1.

existence of triangles similar to a given triangle that was assumed by Wallis. From this assumption he derived the parallel postulate. The argument is similar to Arabian ones, and Wallis actually commissioned a Latin translation of Nasir Eddin al-Tusi, to which he referred in the second of his two lectures given on the evening of 11 July 1663.[1]

Take any two points A and B on a straight line c, and lines a through B and b through A making angles α and β as shown, such that $\alpha + \beta < 2R$. We shall construct a point C on a and b thereby showing that they meet. Draw through B a line b' enclosing an angle α. Slide b along c until it overlaps b', keeping it all the time at an angle α to c. Clearly at some point it meets a; we exhibit one such line b_1 and call the appropriate points on c and a A_1 and C_1 respectively. Now, by assumption, we can draw a triangle ABC on AB similar to A_1BC_1, and then C must lie on a and b as required.

On the face of it this is a direct proof of Euclid's fifth postulate, formulated as it was in the original, and so it is, if we are willing to incorporate Wallis's assumption as an axiom. However, the concept of form independent of size is no more self-evident than the concept of parallels, and we must accept a weaker conclusion: the existence of similar triangles is equivalent to the existence of parallel lines. In other words, if there is to be a geometry in which the parallel postulate fails, then it can contain no figures of the same shape but arbitrarily different sizes. We cannot shrink or expand without distortion.

Giordano Vitale (1633–1711)

Vitale began in the classical fashion by defining parallels as equidistant lines.[2] We may concede the existence of a locus which is everywhere equidistant from a straight line but dispute whether it is itself a straight line. Vitale attempted to prove that it was and failed, giving only a mistaken proof. Unfortunately his mistake is not interesting to us, but the way he succeeded in formulating the problem is. If the equidistant curve to a straight line is not itself straight, pairs

---

[1] Wallis, *De postulato quinto, et definitione quinta, Lib. 6, Euclidis, disceptatio geometrica, Opera Math.*, Vol. II, 669–78 (1693).
[2] Vitale, *Euclides restituto Libri XV*, Rome (1690).

of straight lines must presumably be either concave up or concave down, neither of which possibilities is admittedly very plausible. In order to determine which, if either, is really conceivable, we might proceed as follows.

On the straight line *l* take two points A and B and draw equal perpendiculars BC and AD above each one. There is a unique straight line through C and D, which we draw and call CD, which either behaves as we would expect or is aberrant in the fashion of Fig. 4.2(a) or Fig. 4.2(b). Vitale proved that, if K is any point on AB and H is the point on CD perpendicularly above K (i.e. with the right angle at K), (i) in every case the angles at C and D are equal and (ii) if ever HK = AD then the angles at C and D are right angles and DC is equidistant from AB.

It is the second result which is richer; Bonola called the two results together 'a most remarkable theorem'. In order to prove the existence of parallels on Vitale's definition it would now be enough to prove the existence of *one* point K on AB such that KH = AD. In such a situation it is often possible to show the existence of such a point on general grounds without explicitly locating it.

Therefore, in the case of Vitale's theorem, we might hope that the concepts of distances or straightness might, if elucidated, guarantee us the point K, and once we have K we have a quadrilateral whose angle sum is 360°, as is always the case in Euclidean geometry. We might now advance upon our goal in this fashion: prove that every quadrilateral in our geometry has an angle sum of 360°, then that every triangle has an angle sum of 180°, and finally that the parallel postulate holds. These conjectures would veritably checkmate the opposition. Of course, they would need to be proved, but we would seem to be in a strong position.

It was not to be. Vitale failed to prove the existence of the point H and with it the parallelism of AB and DC. His figure returns in later work, and with it the tantalizing worlds in which the parallel postulate is denied. To glimpse what they are like, consider not the elusive straight line equidistant from a given one, but, as in our original argument, the lines AB and DC. DC must be either concave as in Fig. 4.2(a) or else convex as in Fig. 4.2(b).

In the concave case points H between D and C satisfy HK < CB and the line CD when produced moves away from AB. In the convex case HK always exceeds CB and CD would perhaps approach AB asymptotically or, more excitingly, hit it. Clearly the convex case is more promising if we aim to

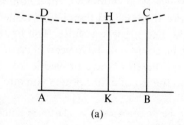 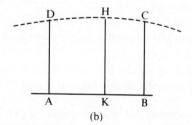

Fig. 4.2.

establish the validity of the parallel postulate by showing that the alternatives are self-contradictory. Equally clearly, these thoughts are no more than suggestive, and much work would have to be done to turn them into proofs of anything.

Gerolamo Saccheri (1667–1733)

The most fullblooded attempt upon the parallel postulate during the eighteenth century was that of Gerolamo Saccheri, who published his *Euclid freed from all imperfections* in Milan in 1733. In it he dealt with the postulate and claimed in the end to have established it. He was, in a curious fashion, correct half of the time.

Saccheri was born in San Remo in 1667 and became a Jesuit in 1685, and it was a Jesuit Father, Tommaso Ceva, who introduced him to Euclid. (Tommaso's brother Giovanni, who was also a friend of Saccheri's, is the man after whom Ceva's Theorem in geometry is named.) Saccheri's first interest was in logic, particularly as exemplified in Euclid, and he esteemed above all arguments that of *reductio ad absurdum*. His only other book is on logic, the *Logica demonstrativa* of 1697 and 1701. When he died in 1733 he was Professor of Mathematics at the University of Pavia.

Saccheri shares with Sir Henry Saville—sometimes called the 'zeroth' Savilian Professor because of the Chairs he endowed at Oxford in the early seventeenth century—a belief that Euclid has two blots. Saville wrote in his *Lectures on Euclid's elements* (1621): 'In the most fair frame of geometry there are two defects, two blots' (quoted by Heath (1956, vol. I, p. 105)). Saccheri chose not only the same word (*naevi*, blot) but the same topics: parallels and the theory of proportion. It does not follow that he knew of his illustrious predecessor, and he nowhere mentions him by name; the same defects had been picked up by Arab writers much earlier.

I propose to summarize Saccheri's argument as it is given in Halsted's translation of 1920, which is a 'parallel text' edition. In so doing I shall be able to employ the simplifications suggested by Bonola in his chapter on Saccheri.

Unlike previous workers, who aimed to establish the existence and uniqueness of parallels as a theorem directly deducible from the rest of geometry, Saccheri aimed negatively to prove that denying the postulate leads to a contradiction. He assumed the first 28 propositions of Euclid[3] and then assumed further that the parallel postulate was false, hoping to find some proposition which was both true and false. For there would then be only one way out of such a mess: some assumption he had made was false. The first 28 propositions of Euclid being unassailable, he could only have been in error in denying the parallel postulate and he would therefore have proved it true. The power of this method is that by adding to the list of assumptions you add

---

[3] Although Euclid introduces parallels into I, 27, he does not need the postulate until I, 29, so the first 28 propositions are independent of it (see p. 37).

to the range of arguments available. As we shall see, the negations of the parallel postulate can be formulated quite precisely.

Saccheri's first achievement was to do that. He worked with a quadrilateral ABCD with right angles at A and B, as did Vitale, whose name he does not mention (see Carruccio 1964, Chap. XV). Let us call the angles at C and D γ and δ respectively. Saccheri proved that

(i) if AD = BC then $\gamma = \delta$.

As a consequence of (i), if AD = BC we may consider three possibilities. Either

$$\gamma = \delta = 90°$$

or

$$\gamma = \delta > 90°$$

or

$$\gamma = \delta < 90°$$

and each of the last two deny the parallel postulate. Saccheri's names for these possibilities were the hypothesis of (i) the right angle, (ii) the obtuse angle, and (iii) the acute angle, in that order (abbreviated to (i) HRA, (ii) HOA and (iii) HAA). Each hypothesis is assumed to hold in the first instance in one given quadrilateral only, but not necessarily in any others. We are at this stage still free to imagine an obtuse-angled quadrilateral in one region of the plane and an acute-angled quadrilateral somewhere else.

With a view to tidying up the situation Saccheri first considered the horizontal arms of the quadrilateral. On the HRA he found, of course,

AB = DC

On the HOA he found

AB > DC

and by a similar argument, on the HAA he found

AB < DC

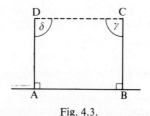

Fig. 4.3.

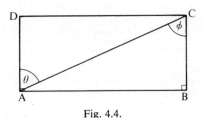

Fig. 4.4.

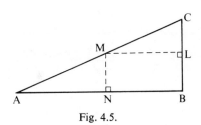

Fig. 4.5.

the inequalities being everywhere reversed. (To prove any one of these split the quadrilateral with a line O'O drawn perpendicularly through the midpoints of DC and AB, and view each quadrilateral sideways, using Vitale's result (ii).)

He was now able to prove a 'three musketeers' theorem: if one of the three hypotheses holds for at least one quadrilateral, then it holds for all. This greatly simplifies the situation and eliminates the chaos we were prepared to tolerate earlier. His proof is tedious, and we shall have occasion later to give a simpler one for the exactly analogous problem with triangles (p. 71), so we omit it. The theorem proves that on each hypothesis space is homogeneous, i.e. it is geometrically the same everywhere.

As far as our earlier pictures are concerned, we may regard the concave picture as being appropriate to the HAA, and the convex one as appropriate to the HOA.

Saccheri next proved the following three-part result (Prop. IX): On the HRA, HOA or HAA the angle sum of a triangle is 180°, >180°, or <180° respectively, and this we shall prove.

First take a right-angled triangle ABC and enclose it in a quadrilateral ABCD such that $D\hat{A}B = 90° = \hat{B}$ and $DA = CB$. On, say, the HOA $AB > DC$, and so $A\hat{C}B > D\hat{A}C$.[4] Since $D\hat{A}C + C\hat{A}B + \hat{B} = 180°$ we must have $A\hat{C}B + C\hat{A}B + \hat{B} > 180°$ which is the desired result for a right-angled triangle. The result for a general triangle follows by dropping a perpendicular from a vertex and splitting it into two right-angled triangles.

The proof of the theorem for the other two hypotheses is similar.

[4] This is not obvious. However, we know from Euclid I, 18, which is not in dispute, that smaller angles are opposite smaller sides. Since $DC < AB$, $\theta < \phi$ or $D\hat{A}C < A\hat{C}B$.

The HOA discussed

Take now the HOA. Let ABC be a triangle with a right angle at B and let M be the midpoint of AC. Drop the perpendicular MN from M to AB. We shall show that NB > AN (Saccheri, Prop. XII).

We have $M\hat{C}B + C\hat{B}N + B\hat{N}M + N\hat{M}C > 360°$ and so $M\hat{C}B + N\hat{M}C > 180°$, but $A\hat{M}N + N\hat{M}C = 180°$, and so $M\hat{C}B > A\hat{M}N$. Drop the perpendicular ML from M to BC. AMN and MCL are right-angled triangles with equal hypotenuses, so our result above implies that ML > AN. However, NML > 90°, since the other angles in LBNM are all 90°, and so NB > ML and NB > AN, as was required to be proved.

Thus, if equal intervals are taken along a sloping line AC, these project vertically into increasing intervals along a horizontal line AB.

We may now refute the hypothesis of the obtuse angle yet again. Let AB and CD be two lines cut by AC and such that $B\hat{A}C + A\hat{C}D < 180°$. Then one of these two angles must be acute, $C\hat{A}B$ say. Drop the perpendicular CH from C to AB. Then $A\hat{C}H + C\hat{H}A + H\hat{A}C > 180°$ by the HOA. By assumption, however, $B\hat{A}C + A\hat{C}D < 180°$. Therefore $C\hat{H}A > H\hat{C}D$, i.e. $H\hat{C}D$ is acute.

We now prove that AB and CD must meet, thereby establishing the parallel postulate which we assumed to be false! Therefore it must after all be true, in accordance with the method of arguing established at the beginning.

CD makes an acute angle with CH and HB is perpendicular to it. Draw them with BH vertical. Take M_1 on CD and drop the perpendicular M_1N_1 onto CH, and then repeat with M_2 on CD and N_2 on CH_1 chosen such that $2CM_1 = CM_2$. We know that $CN_1 < N_1N_2$ and so $CN_2 > 2CN_1$. Carry on the process indefinitely, choosing M_{n+1} such that $CM_n = M_nM_{n+1}$, i.e. $2_nCM_1 = CM_{n+1}$, and we have $CN_{n+1} > 2^nCN_1$, and the points N_n move indefinitely far from C. However, CH is a fixed distance and so eventually we can choose N_k such that $CN_k > CH$ by choosing k such that $CN_k > 2^{k-1}CN > CH$. Therefore H lies inside triangle CN_kM_k, and HB must therefore meet CM_k, i.e. it must meet CD. Thus HOA is refuted.

A proof freed from all pedantry is as follows. The M's move out along CD, and the N's along CH. The further out M is, the further N is without limit.

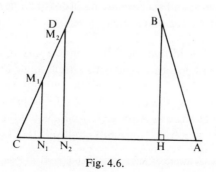

Fig. 4.6.

Therefore N eventually lies beyond H. Locate the intermediate point where it overlapped.

Saccheri remarked (Prop. XIV): 'The HOA is false because it destroys itself.'

TWO REMARKS. It may seem perverse to establish that two lines, one perpendicular to a third and one inclined acutely to it, must meet, but it was necessary to do so since, after all, it is an assertion strikingly similar to the parallel postulate itself. More significantly, however, Saccheri assumed in proving it that a line segment may be extended indefinitely,[5] i.e. that straight lines are infinite in extent. We could better formulate our final result this way. On the assumption that straight lines are infinite, the HOA is self-contradictory (see p. 149). As Bonola says, the use of the theorem of the exterior angle (Euclid I, 16) is a tacit assumption of the infinitude of lines.

Interestingly enough, we find Aristotle saying, in his *Posterior analytics* (I, 5), that it can be deduced that parallels meet from the assumption that the angle sum of a triangle exceeds two right angles (Heath, 1949, p. 41).

The HAA

Saccheri: 'And here begins a lengthy battle against the HAA which alone opposes the truth of the axiom'.

There remains the HAA. The obvious approach to take is to mimic the arguments used in the case of the HOA, but this fails. When we drop perpendiculars from one on to another, as again in Fig. 4.6, the equal intercepts $AM_1 = M_1M_2 = M_2M_3 = \cdots$ project into steadily smaller intercepts

$$AN_1 > N_1N_2 > N_2N_3 > \cdots$$

so that the argument used to conclude the proof in the HOA case fails utterly. We cannot guarantee that there is any N_k reached such that $CN_k > CH$. Nor can a subtler argument be found, which allows for the intercepts N_k to crowd together but shows that none the less they finally reach far enough, like the paces of a weary man.

If the line and the perpendicular do not meet what does the world look like?

Saccheri proved that, on the HAA, you can find lines *l*, *m*, *n* with *l* meeting *m* at an acute angle and *n* perpendicular to *m*, but such that *l* and *n* do not meet. The construction and accompanying proof are a neat illustration of the new clothes Euclidean theorems must have put upon them in this strange world.

We known from Euclid that the following is a way to construct lines that do not meet. Let ABC be a triangle with a right angle at B. At C draw CD such that $D\hat{C}A = C\hat{A}B$. Then CD and BA do not meet, as Euclid I, 27 asserts; the result is also true on the HAA. (If CD produced met BA, at X say, then the external angle at A to XAC would equal the internal angle at C and so $B\hat{A}C < A\hat{C}D + A\hat{X}C$, but this contradicts Euclid I, 17.) This is

---

[5] Saccheri was somewhat aware of this; see his scholion II to Prop XIII.

independent of any business about parallels, in particular; on the HAA it remains true, but on the HAA the angle DCB is acute (the angle sum of a triangle is less than 180°) and so the conditions of Saccheri's theorem obtain.

Common perpendiculars

With regard to a pair of lines which do not meet, a and b say, we can look for a common perpendicular.[6] This would measure the minimum distance they are apart. Take A_1 and A_2 arbitrarily on a and drop perpendiculars from there to b as shown. In the quadrilateral $A_1 A_2 B_2 B_1$ either the interior angles A_1 and A_2 can be (i) both acute, or (ii) one right and one acute, or (iii) one acute and one obtuse.

In the first case there must be a common perpendicular to a and b between A_1 and A_2, as can be seen by watching the angle at A_1 as you slide $A_1 B_1$ along to A_2.

In the second case we have already accidentally drawn the common perpendicular.

In the third case there is no common perpendicular between A_1 and A_2, nor will there ever be if the angle at A_2 is always obtuse wherever A_2 lies to the right of A on a.

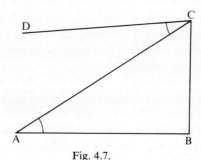

Fig. 4.7.

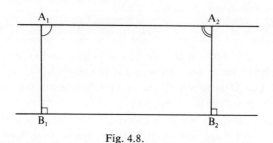

Fig. 4.8.

[6] There can only be one. Here Saccheri discussed the work of Nasir Eddin and Wallis, which he showed involves the equidistance of lines.

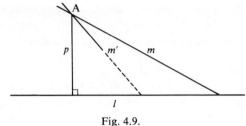

Fig. 4.9.

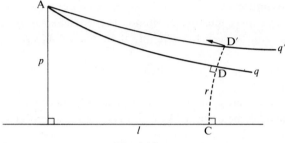

Fig. 4.10.

As to non-intersecting lines with no common perpendicular Saccheri showed that they must approach each other asymptotically, and his arguments in Prop. XXIII introduce a figure of fundamental importance to all subsequent work.

Consider a point A not on a line l. The lines through A divide into two groups, those which meet l and those which do not meet l, and of those some have a common perpendicular with l and some do not. If we draw p, the perpendicular to l through A, then we encounter these groups successively since (i) lines such as m', which makes a smaller angle with p than a line m meeting l, must themselves meet l and (ii) any line q' making a larger but still acute angle with p than q, a line having a common perpendicular with l, must also have a common perpendicular with l.

The common perpendicular is shown dotted between q and l; one must be found between q' and l. Indeed, let r be the common perpendicular to l and q, and let it meet l at C and q at D. Extend r to meet q' and D' say. The perpendicular to r at D' must pass north of A and so the angle AD'C is acute. Therefore quadrilateral ABCD' is of type (i) above, and there is a common perpendicular to l and q' which in fact lies between p and r.

Given a line r through A which meets l, we may easily find a line r' lying above it which still meets l. Indeed, it is enough to join A to any point of l to the right of where r meets l. Therefore, if we let r swing around anticlockwise, making steadily larger angles with p at A, we never encounter a last line of the sort which meets l. The significance of the lines through A which have a common perpendicular with l is that they prescribe an upper limit to the angle between p and r for lines r which meet l. Call this upper limit α.

(iii) Had we begun with r at right angles to p, in which position it certainly does not meet l, and rotated it clockwise a similar thing would have happened. No last position could be found for r in which it had a common perpendicular to l but no lines below it did. Therefore there must be a lower limit β never attained for the angle between r and p for lines r having a common perpendicular with l (Saccheri, Prop. XXX).

Evidently $\alpha < \beta$. What of the lines r_α, r_β making angles with p at A of α and β respectively? A plausible guess would be that they are the same line, and indeed they are (Saccheri, Prop. XXXII). (To prove that r_α is r_β suppose that it is not and consider any of the lines that there would then be in between.) This line is asymptotic to l. We may summarize the work to date as follows.

On the HAA the family of lines through a point A not on a line l contains two lines r and r', one asymptotic to l to the right and the other asymptotic to l to the left, dividing the pencil into two parts. The first part contains those lines which meet l, and the second part those lines which have a common perpendicular with l.

At this stage we must part company with Saccheri. He said:

At length I have disproved the hostile HAA by a manifest falsity, since it must lead to the recognition of two straight lines which at one and the same point have in the same plane a common perpendicular. (Halsted edn, pp. 14, 15.)

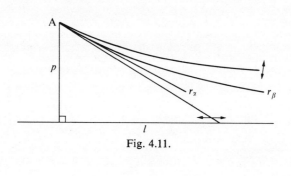

Fig. 4.11.

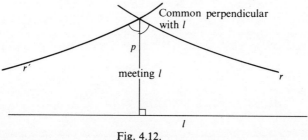

Fig. 4.12.

However, his 'proof' of the impossibility of the HAA is invalid, extending to infinity as it does certain properties valid for figures at a finite distance, in particular the idea that two lines might have a common perpendicular at the point at infinity where they meet. The language of points, lines, etc. 'at infinity' is never clear and is often meaningless, and Saccheri was by no means the last person to be misled by it. However, note that if two lines have a common perpendicular at a common point not at infinity they turn out to be the same line.

The hypothesis then remains intact, surviving the last and most powerful of the classical attempts upon it. Subsequent work, as we shall see, casts this vindication of Euclid unjustly into the shadows, whereas it represents a considerable clarification of the problem. Two alternatives to the parallel postulate have been formulated and shown to be the only possible ones. One is shown to contradict itself, upon admitting a plausible assumption about straight lines, and the other to lead, tantalizingly, to the figure above. The failure to refute the possibility of such a geometry, in the face of these efforts, must have suggested the idea that, indeed, such a new geometry was possible.

5 J. H. Lambert's work

It is a commonplace in the history of science that navigation was the source of many important problems. The expanding mercantile world needed increasingly accurate methods of determining the location of ships at sea and of measuring the passage of time, and the science of the day was obsessed with these issues. In navigation it is a commonplace that the shortest line between two points is the segment of the great circle between them. A great circle on a sphere is a circle of maximal radius and is made by cutting the sphere with any plane circle which passes through its centre. Thus the lines of longitude are great circles, but none of the lines of latitude are, except for the equator.

A puzzling 'geometry'

What can be said about the geometry of great circles on the sphere? It is in many respects like our own. Two points determine a unique great circle which joins them, unless they are antipodal, and we may draw triangles and circles. Of course, two great circles meet not in one point but two, the second point diametrically opposite the first, but this is an inconvenience we might easily talk ourselves round. More disturbingly, all great circles meet, and so there is no possibility of defining parallelism for this geometry. If we then consider a triangle we notice that its angle sum is in excess of 180°. Furthermore, we cannot take a triangle and shrink it onto a smaller but similar one. As we shrink the triangle its angles change, and indeed become smaller. No great circle can be added to a triangle of great circles which makes the corresponding base angles equal. We have on the sphere a geometry strikingly like the

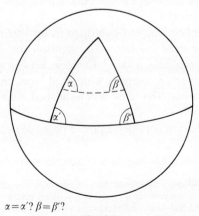

$\alpha = \alpha'$? $\beta = \beta'$?

Fig. 5.1.

geometry obtained on the (refuted) HOA. It is obstinately and demonstrably 'there' and was known to the mathematicians of the day, who, however, chose to regard it as a geometry on the sphere and not on the plane. Therefore it was amusing but not relevant to the problem of parallels. It could be, and was, ignored, and the refutation allowed to stand. Yet it haunted all subsequent workers in the field, and we shall return to it more than once.

Strictly, spherical geometry differs from Euclidean in two respects, the non-existence of parallels and the finitude of lines, so the refutations of the HOA are logically correct. However, Lambert and Taurinus none the less accepted spherical geometry as an example of HOA geometry, as we shall see, so for them the question was not merely a logical one but one concerning the geometry of space.

Johann Heinrich Lambert (1728–1777)

The next mathematician of interest in the study of the problem of parallels is the Swiss mathematician Johann Lambert. He was a wide-ranging thinker, one of the first to consider our galaxy finite and only one island universe out of many—a distinction he shares with Kant (see Lambert's *Cosmologische briefe*† (1761)) and he deserves more attention than he has had. He was an early member of the Berlin Academy of Sciences with Euler, and contributed to many areas of knowledge: to photometry and optics (1760), where he established Lambert's cosine law for the intensity of a refracted beam of light as well as the inverse square law for the intensity of illumination at a distance; to statistics; to pure mathematics, where amongst other things he was the first to prove that π was irrational (1766); to logic, where he extended Leibniz's symbolic studies; to astronomy, where he established Lambert's law for the parabolic orbit of a comet (1761). In his *Freye perspektive* (1759, 1774) he not only discussed perspective drawing but also the intriguing problem of what geometric constructions can be made with a straight edge alone and only a fixed compass. Like Kant he sought to reintroduce the *a priori* into science, and he wrote purely philosophical works which now meet with differing assessments of their merits.

His *Theory of parallels* was not published in his lifetime, unlike Saccheri's work which had attracted considerable attention for a short while and of which Lambert was probably aware.[1] Lambert was possibly unsatisfied with his own study, which often happens with good work that cannot be carried through to its conclusion. The book was published after his death by Johann Bernoulli III in 1786.

He followed Saccheri in considering three possible values for the angle sum

---

† Translated as *Cosmological Letters*, by S. L. Jaki, Scottish Academic Press, Edinburgh, 1976.

[1] Lambert refers to Klügel's historical thesis *Conatuum praecipuorum theoriam demonstrandi recensio, quam publico examini* ... of 1763, written under the direction of Kästner whom Lambert knew. Klügel discusses Saccheri's work; Lambert in turn mentions Klügel by name but not Saccheri.

of a quadrilateral, and like him rejected the HOA but was unable to reject the HAA. In the course of his researches he made two interesting observations. In Euclidean geometry there is an essential difference between the measurement of angles and lines. There is an absolute measure for angles, since the whole angle (360°) is of a fixed size however it is drawn. No such situation prevails for lines, however, and we must arbitrarily define the length of some segment before we can make any measurements.

Thus the angle at A (see Fig. 5.2) is a definite proportion of the whole angle at A and its size can be unambiguously determined as the appropriate ratio. However, the segment AB varies in length according to the segment taken as defining the unit of length; it may be anything (1 cm, 1 km, 1 mile, 1 μm, etc.).

The absolute nature of length

On the HAA Lambert was the first to notice that this ambiguity disappears; it is possible to define an absolute unit of length. To do so, it is enough to associate one line segment uniquely with one angle and thereby transfer the absolute measure for angles onto lengths as well. Let us take an angle of, say, 50°. There will be an equilateral triangle whose angles are each 50° since $50° + 50° + 50° = 150° < 180° = 2R$ and any other triangle will be congruent to it. However, there are none merely similar to it by Wallis's result, so we have associated a unique length to an angle of 50°. In this way, length becomes absolute in the new geometry. (Attempting to repeat the proof in the Euclidean case fails precisely because similar triangles exist there.) For the sake of convenience we should like this definition of distance to be additive which, as it stands, it is not.

On this definition (see Fig. 5.3) the distance from A to C is shorter than the

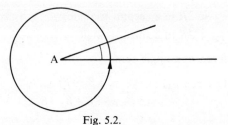

Fig. 5.2.

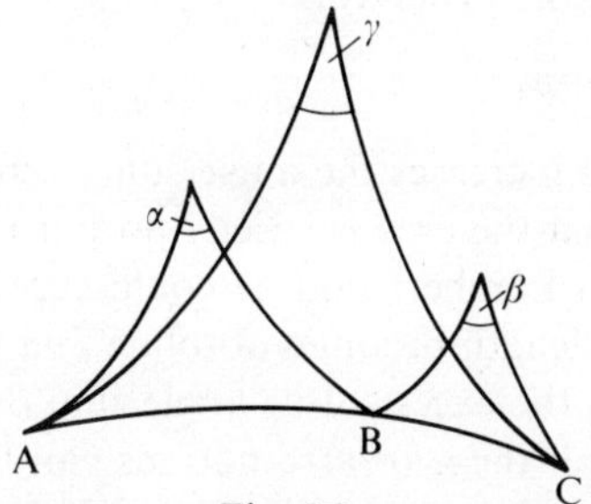

Fig. 5.3.

66   J. H. Lambert's work

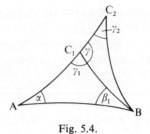

Fig. 5.4.

distance AB plus the distance BC. However, this can be rectified once we define the distance AB to be some specified function of the appropriate angle. The function concerned is a trigonometric one which we shall meet later on.

This absolute measurement of length is repugnant to our Euclidean intuition, and Lambert seems to have wished to throw out the new geometry on this account. However, he wisely refused to consider that he had done so.

Angle sum and area

Lambert also noticed that the difference between the angle sum of a triangle and 180°, on the HAA, decreases with area, as can be seen by adding triangles together.

Triangle C_1AB has had its side AC_1 extended to C_2 and a second triangle AC_2B drawn. In each case the angles are $\alpha, \beta_1, \gamma_1$ and $\alpha, \beta_2, \gamma_2$ respectively. Evidently the area of $\triangle ABC_2$ exceeds $\triangle ABC_1$:

$$\alpha+\beta_1+\gamma_1 < 2R \text{ by hypothesis in one triangle}$$
$$\gamma_1+\gamma = 2R$$
$$\gamma+\beta_2-\beta_1+\gamma_2 < 2R \text{ in a second triangle}$$
$$\alpha+\beta_2+\gamma_2 < 2R \text{ in a third triangle.}$$

Therefore the difference in angle sums

$$\alpha+\gamma_2+\beta_2-(\alpha+\gamma_1+\beta_1) = \gamma_2-\gamma_1+\beta_2-\beta_1$$
$$= \gamma_2+\gamma-2R+\beta_2-\beta_1$$
$$< 0$$

as the second triangle shows. Therefore

$$\alpha+\gamma_2+\beta_2 < \alpha+\gamma_1+\beta_1$$

and therefore as the area increases the angle sum decreases. It can be shown by a subtler argument that the area is strictly proportional to $2R-(\alpha+\beta+\gamma)$ (see the exercises), which Lambert said he could supply but did not (§ 81).

Likewise on the HOA length becomes absolute and we may relate the area of a triangle this time to the excess of its angle sum over $2R$.

In fact, analogues of all these bizarre notions can be illustrated easily on spherical geometry, as Lambert said (§ 82). On a sphere a segment AB defines an equilateral triangle ABC with angle α, say. Notice 3α > 180°. Since the only similar triangles to ABC are congruent to it we may start with an angle α and reconstruct ABC uniquely. Furthermore, as Girard[2] knew, the angle sum of a triangle on a sphere determines its area. On a sphere of radius r a triangle with angles $\hat{A}$, $\hat{B}$, and $\hat{C}$ has area $r^2(\hat{A}+\hat{B}+\hat{C}-\pi)$, so any attempt to shrink a triangle onto a smaller one must necessarily reduce the angle sum.

Curiously, there is an observation of some profundity which does not seem to have been made by the workers at this time. Just as the absolute measure for angles is consequent upon the existence of an angle of maximum size—the whole angle—there is also a segment of maximal length, namely the great circle. I shall say no more at this stage, but if you try to re-work the refutation of the HOA in the setting of spherical geometry you will find this fact obstinately relevant.

Lambert's attitude to spherical geometry may be gauged by the following remark. Noticing that if the radius of the sphere taken is r, then the area of a triangle with angles α, β, and γ is $r^2(\alpha+\beta+\gamma-\pi)$ and that in the case of the HAA formulae for an area like $r^2\{\pi-(\alpha+\beta+\gamma)\}$ occur he remarked (§82):

From this I should almost conclude that the third hypothesis would occur in the case of the imaginary sphere.

He was thinking that on a sphere of radius $(\sqrt{-1})r$ the formula becomes

$$\{(\sqrt{-1})r\}^2(\alpha+\beta+\gamma-\pi) = (-r^2)(\alpha+\beta+\gamma-\pi) = r^2\{\pi-(\alpha+\beta+\gamma)\}.$$

This temerity has cost Lambert the popular fame accorded to the discoverers of non-Euclidean geometry. I believe differently that it marks him as a correct and inspired thinker. The notion of a sphere of imaginary radius is quite unclear, and the reference to geometry on a sphere could not be properly articulated anyway until much later on. To advance as far as knowledge permits without disguising conjecture as discovery is the business of research, and as a conjecture Lambert's remark is provocative beyond all others. To enter the land which Lambert's vision was the first to descry was to take mathematics another hundred years.

Exercises

5.1 Show that on a sphere of radius 1 the area of a triangle with angles $\hat{A}$, $\hat{B}$, $\hat{C}$ is $\hat{A}+\hat{B}+\hat{C}-\pi$ by extending the sides into complete circles and considering the lines so obtained.

The total area of the sphere is 4π. The area of the two lines at $\hat{A}$ is $4\hat{A}$, etc. The sum of the areas of the three (double) lunes is the whole sphere plus triangle ABC counted four times (why?). Therefore

$$4(\hat{A}+\hat{B}+\hat{C}) = 4\pi + 4(\text{area ABC})$$

whence the result.

---

[2] Albert Girard (1595–1632); see Coxeter (1961, p. 85).

68   J. H. Lambert's work

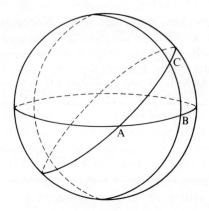

Fig. 5.5.

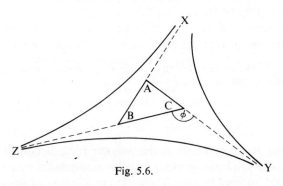

Fig. 5.6.

5.2 Complete the steps in the following argument, which is originally due to Gauss,[3] and thereby show that in non-Euclidean geometry a triangle with angles $\hat{A}$, $\hat{B}$, $\hat{C}$ has area proportional to $\pi - (\hat{A} + \hat{B} + \hat{C})$.

(1) Assume any figure made up of three mutally asymptotic lines has the same finite area t, say.
(2) Consider the area of a figure such as ZCY as a function f of $(\pi - \phi)$—Gauss wrote $180° - \phi$.
(3) Deduce from a suitable figure that $f(\pi - \phi) + f(\phi) = t$.
(4) From another figure deduce that $f(\phi) + f(\psi) + f(\pi - \phi - \psi) = t$.
(5) Deduce that f is additive, $f(\phi) + f(\psi) = f(\phi + \psi)$, and therefore that $f(\phi)/\phi = $ constant.
(6) Finally, deduce that area $(ABC) = t\{\pi - (\hat{A} + \hat{B} + \hat{C})\}/\pi$.

An elementary proof, due to H. Liebmann, that the assumption that t is a finite constant is not necessary, but can be proved as a theorem in non-Euclidean geometry, can be found in Coxeter (1961) p. 295. It is easily proved using the methods of differential geometry, see below p. 123.

[3] Letter to Farkas Bolyai, on receipt of the *Tentamen*, 6 March 1832 (*Werke* VIII, pp. 220–4).

6 Legendre's work

The powerful French school of mathematics contributed to every branch of the subject; indeed towards the end of the eighteenth century, after the death of Euler (1783), they dominated it entirely. The only problem which seems not to have interested them significantly was the problem of parallels. Their contributions can be summarized quite briefly, following Bonola, p. 54. Laplace in 1824 argued that the principle of similarity is inherent in nature and that we cannot conceive of the universe as absolute in size. Instead we can increase or decrease the size or objects in proportion without changing, for instance, the nature of orbits of the planets; therefore, he said, space itself has this property of admitting similar figures. On that assumption the usual theory of parallels follows. Lagrange, we are told, held that spherical trigonometry was independent of Euclidean geometry and was true whatever you believe about parallels—a result which was to be important in the subsequent development of the subject. However, he does not appear to have made much use of this observation himself. Fourier, taking his cue from an earlier remark of D'Alembert, believed that the answer to the problem lay in having a good definition of a straight line, which he connected with the idea of distance, but once again nothing seems to have come of this. I might add that Poncelet,† in his *Traité des propriétés projectives des figures* (1822), concentrated on those properties of figures which are unaltered by projection and seems not to have considered the nature of the ambient space. Their lack of interest in the problem is remarkable, and I shall discuss it later.

Only Legendre[1] made any comprehensive attempt upon the problem, and that in a conservative fashion, hoping to prove that the parallel postulate is valid as a theorem in geometry. He frequently proved theorems known earlier to Saccheri, but his proofs have the merit of elegance and like Bonola I have preferred them for that reason. Legendre followed Saccheri, whose work, however, he may not have known, in starting with hypotheses about the angle sum of a figure, in this case the triangle. His results have come to be known as the theorems of Legendre, although they were known to others before him. In his so-called *first theorem* he established that the angle sum of a triangle cannot exceed $2R$ (*Eléments de géométrie*, 2nd edn, Prop. 19). The proof is as follows.

Take n equal segments $A_1A_2, A_2A_3, \ldots, A_nA_{n+1}$ on a line and construct a row of congruent triangles upon them as shown. Join up the B's to obtain a

---

† The science of projective geometry begun by Poncelet will not be discussed here, but see the books by Coxeter and Pedoe in the Bibliography.
[1] *Eléments de géométrie* (1794–1823), with more editions after his death in 1833.

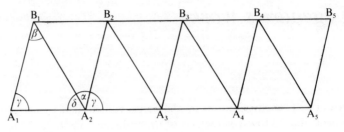

Fig. 6.1.

second set of congruent triangles, pointing downwards, and add B_{n+1} so that $B_n A_{n+1} B_{n+1}$ is also congruent to the downward triangles. Let the apex angle of the upward triangles be β in each case, and the apex angle of the downward ones be α. Without parallels we cannot assume $\alpha = \beta$. Looking carefully at the start we see the following equality. The angles lower left, $B_1 \hat{A}_1 A_2$, $B_2 \hat{A}_2 A_3$, etc. are all equal to γ, say. Now at A_2, if $A_1 \hat{A}_2 B_1 = \delta$,

$$\delta + \alpha + \gamma = 180°$$

and

$$\delta + \beta + \gamma = \text{angle sum of } B_1 A_1 A_2$$

so it will be enough to show $\beta \leq \alpha$. Very well, suppose it is not, i.e. $\beta > \alpha$. Then $A_1 A_2 > B_1 B_2$, since $A_1 B_1 = A_2 B_2$ and $B_1 B_2$ is common to both $A_1 B_1 A$ and $B_1 A_2 B_2$, so $A_1 A_2 - B_1 B_2 > 0$ (cf. Euclid I, 18: In any triangle the greater side subtends the greater angle). We also know that

$$A_1 B_1 + B_1 B_2 + \cdots + B_{n+1} A_{n+1} > n A_1 A_2$$

top route A_1 to A_{n+1} > bottom route. Therefore

$$A_1 B_1 + n B_1 B_2 + B_{n+1} A_{n+1} > n A_1 A_2.$$

Therefore

$$2 A_1 B_1 > n(A_1 A_2 - B_1 B_2)$$

since $A_1 B_1 = B_{n+1} A_{n+1}$ which we have just said is positive. However, $A_1 B_1$ is fixed, and n is not. Therefore if we take n large enough we arrive at a contradiction and thus refute $\beta > \alpha$ and thereby establish the theorem.

Not content with this refutation, which is akin to Lambert's, Legendre offered a second.[2] It is fallacious, and spotting the flaw is a disturbing exercise which will be answered later (p. 152).

Let ABC be a triangle with angles α, β, γ such that $\alpha + \beta + \gamma < 180°$; for definiteness let the defect, which is defined to be $180° - (\alpha + \beta + \gamma) = \delta$. Locate A' symmetrically situated to A with respect to BC (this can be done by

---

[2] *Mem Acad. Sci Paris* (1833); *Eléments de géométrie*, Note II, pp. 274, 276 (1823).

rotating ABC through 180° around the midpoint of BC) and extend AB and AC. Draw through A' a line meeting AB at B' and AC at C'—not necessarily a line 'parallel' to BC which would beg the question. Join up A'B and A'C. By symmetry the defect of triangle A'BC is also δ. Since the defect of the angle sum of the large triangles AB'C' is the sum of the defects in the angle sum of the four triangles separately, we have in AB'C' a triangle with defect $\geq 2\delta$. Continuing in this manner we obtain triangles with defects δ, 2δ, 4δ, 8δ and so on, triangle $AB^{(n)}C^{(n)}$ having defect $2^n\delta$. However, for n sufficiently large we have a triangle whose defect is greater than 180°. This is evidently impossible, and so we have proved that the angle sum of a triangle is 180°, from which the ordinary theory of parallels follows.

Alas, this is not so. The proof is flawed, and not this time by appeal to the postulate of Archimedes which permits the indefinite replacement of segments and thereby the '2^n for large enough n' arguments. In spotting the flaw you will discover more about the alien nature of non-Euclidean geometry than by following any texts.

However, on the basis of Legendre's work (and assuming that the HOA has been refuted) we can now give a neat proof of the 'three musketeers' theorem that if the angle sum of one triangle is less than 180° then it is less than 180° in all triangles. This proof is in fact due to Lobachevskii (1840, §20) (see also p. 56). We first prove the corresponding result in the Euclidean case: if in one triangle the angle sum is 180° then it is so in every triangle. Let ABC be a triangle with angle sum $\alpha+\beta+\gamma = 180°$. Then at least two angles, α and γ say, must be acute. Drop a perpendicular p from B to AC. Then the angle sum of $\triangle$ ABD $\leq 180°$ and the angle sum of $\triangle$ DBC $\leq 180°$, but the angle sum of $\triangle$ ABC is 180°. Therefore equality must obtain and the angle sums of $\triangle$ ABD and $\triangle$ DBC are also 180°. BDC is a right-angled triangle with angle sum 180° and sides p and q say. By 'flipping it over' we construct a tile with sides p and q, opposite sides equal, and adjacent sides meeting at right angles. By 'tiling' we can build up arbitrary rectangles with sides np and mq for any m and n, which we can split diagonally into two right-angled

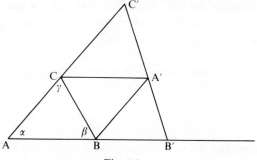

Fig. 6.2.

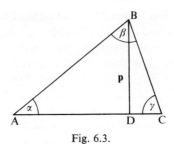

Fig. 6.3.

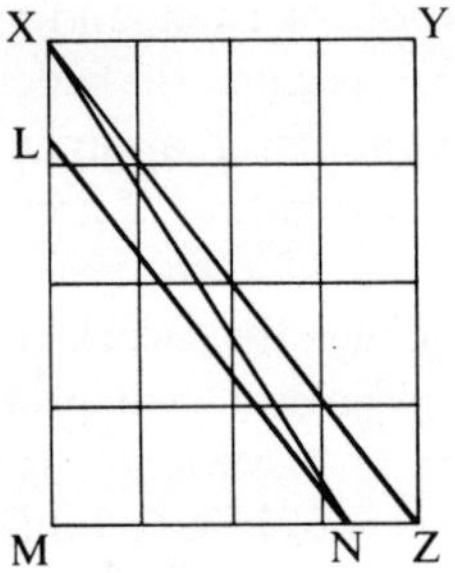

Fig. 6.4.

triangles with angle sum 180°. Now let LMN be a triangle with a right angle at M but whose angle sum is not known to be 180°. We shall prove that it must be. Start tiling with the *pq* tile at M until you have entirely covered △ LMN in a large rectangle:

$$\triangle \text{XMZ} = \triangle \text{XMN} + \triangle \text{NXZ}.$$

Therefore, using the same method of proof as before, the angle sum of both △ XMN and △ NXZ must be 180°. Now △ XMN = △ XLN + △ LMN, so again the angle sum of the two smaller triangles must be 180°. In particular we have proved that the angle sum of △ LMN, and therefore of any right angled triangle, must be 180°. However, this finishes the proof, since any triangle is the sum of two right-angled triangles. If, finally, we know that no triangle has an angle sum greater than 180°, then either (i) one triangle has an angle sum of 180°, in which case all do, or (ii) no triangle has angle sum 180°, in which case all have angle sum less than 180°, and in particular if one triangle has an angle sum less than 180° then they all must.

Exercises

6.1 Prove that on the HAA a triangle can be constructed having all its angles equal to 50°.

6.2 Prove that on the HAA a triangle can be constructed having angles α, β, and γ provided only that $\alpha+\beta+\gamma < 2R$.

6.3 What 'mistake' is made by Legendre in the first of his theorems quoted? What mistake is made in the second?

These problems are discussed in Chapter 14.

7 Gauss's contribution

The decisive breakthroughs in the study of Euclidean geometry occurred at the start of the nineteenth century. The combined studies of the previous hundred years must have led to a picture like the following in the mind of contemporary mathematicians.

Upon the truth of the parallel postulate depends much of ordinary geometry. To deny the postulate is to assert either that the angles of a triangle always add to more than $2R$ or that they always add up to less. The first hypothesis leads to a contradiction and both lead to the existence of an absolute measure of length, which inclines one to reject them. Upon either hypothesis one must also do without similar triangles, and upon the second hypothesis one must accept lines which never meet and lines which are asymptotic to one another.

To this picture we may add two further details, which may characterize their attitude of mind. It is not enough to feel unhappy with the alternative 'geometries'; one should be able to give a rigorous refutation of them. To do so would entail saying exactly what is wrong with, for instance, asymptotic straight lines, and this is just what they could not do. Legendre's proof that the first hypothesis leads to a contradiction has exactly the character we seek; if you like, a blind man could understand it. However, without appeal to pictorial intuition the asymptotic lines are not absurd, if the absolute nature of length is admitted in spherical geometry the same notion cannot be denounced as self-contradictory when it appears elsewhere. The bewilderment may focus itself on the definition of 'straight', perhaps, and one might ask what is straight about 'straight' lines. Fourier had raised this question in 1795 in a discussion with Monge (see Bonola, p. 54) and the leading German geometer of the period, A. G. Kästner, had written in a letter to Pfaff that there was not even a clear definition of a straight line (2 August 1789, see Engel and Stäckel 1895, p. 140). It would seem that the framework of classical geometry, wherein the notion of 'line' and 'point' are undefined, cannot answer these questions.

However, and this is the second detail, geometric knowledge is often considered to be true knowledge. One may determine the shape of the world by reasoning alone. Since Euclidean Geometry is evidently right, the other(s) must be wrong. A new geometry would define a possible world, but *this* world would remain the true one and so by reasoning alone we would have created a false one. In some way rationality would have led us astray. Efforts to refute the alternative hypotheses must be redoubled.

Today we would not accept the second argument, finding it unpleasingly spare of choices, but one of our historical reasons for doing so is the impact of the discovery of non-Euclidean geometries. We must allow our predecessors at least the plausibility of their point of view; it survives today in nearly all elementary accounts of the subject and we have less excuse.

This informal feeling for what constitutes mathematical truth received a novel emphasis from the work of Immanuel Kant. He has come to be regarded as believing that in geometry lay our true knowledge of the world—knowledge *a priori* in that it could be discovered by the mind alone, but synthetic in that it applied to the world we live in. The synthetic *a priori* has remained contentious ever since, though geometry as a particular source example of it has, of course, fallen into disrepute. Bonola (see pp. 64, 121) explicitly argued that it was Gauss's disagreement with Kant on the nature of geometry that enabled him to make his great achievements. Russell, in several books, derived comic mileage from Kant's dogmatic error. The truth, however, would seem to be more complicated. Kant was not committed to a geometry based upon the parallel postulate, which he was prepared to hold in abeyance. Kant's concern was only to refute geometry on the HOA, in which any two lines enclose an area *as not true of space*, which is unique and therefore must be conceived of as Euclidean. He did not believe that a 'non-Euclidean' geometry was logically impossible, however. Geometry is possible with or without the parallel postulate; there are several theorems independent of it which together form the body of 'absolute geometry', and it may be this that Kant had in mind. He corresponded with Lambert on the matter but the letters are inconclusive (Kant (ed. Schöndörffer) 1972). However, he was more concerned with the interrelation of mathematics and natural science in the sense that mathematics provides a frame which empirical science cannot transgress. Accordingly, the easy dismissals of his ideas are misplaced.

Kant's opinions may have influenced mathematicians working on the problem, although the French seem largely to have held to the naïve view that space somehow had to be Euclidean.

There is a more important consideration, however, if you believe that the postulate will ultimately be proved. By 1800 it had resisted several attacks upon it, and revealed its intimate relationship with several properties in geometry of at least equal intuitive force. Therefore its proof could not be an easy result to obtain. All you can guarantee, should you succeed in proving it, is a sigh of relief from the mathematical community. Ah yes, your colleagues would say, we thought so all along. Worse still, no new consequences would follow from your result. Unless you had thrown up a new rich equivalence, or employed a new technique which could be used elsewhere, your work would be a prodigious dead end. Believing that the postulate is true means that no new vistas confront and no challenging exploration can ensue. There is only the prospect of much hard slogging, and success may even then elude you.

Considerations like these may have influenced the French to ignore the problem. The rich field of analysis was being opened up, the solution to many problems of possibly equal charm and fascination was apparently to hand,

and moreover these built up into an impressive new edifice. The gold was, if not lying on the ground, at least in nuggets to be found only a little further up the valley. The creative young mathematician would more readily find success here than in the deadening climate of classical geometry. Doubtless the problem haunted their minds, but it was irksome rather than attractive and best ignored unless you could not leave such curiosities alone.

Gauss

Karl Friedrich Gauss is sometimes considered, for example by Morris Kline, to be the first discoverer of non-Euclidean geometry. He was, in any case, a formidable mathematician, making his first important contributions in this area when he was 15, and the range and profundity of his work place him among the truly great. He first worked in number theory, function theory, and classical geometry, and in his later work treated surfaces, statistics, probability, and mechanics whilst earning his living as an astronomer. In Gauss and Riemann, Germany gave to the world a tradition which created a new way of seeing. With regard to non-Euclidean geometry, however, the usual claims are perhaps somewhat overstated.

It does seem that Gauss was the first to believe that a non-Euclidean geometry was possible and the attempts to find a contradiction in other geometries were therefore in vain. In 1817 he wrote to Olbers[1]

I come more and more to the view that the necessity of our geometry cannot be proved... Perhaps we shall come to another insight in another life into the nature of space, which is unattainable for us now. But until then one must not rank Geometry with Arithmetic, which is truly *a priori*, but with Mechanics...

He kept up with published books on Euclid's postulate, remarking in one review that hardly a year went by without someone writing a book on the subject. In 1816 he expressed his own doubts about the postulate in a review, but the great care with which he expressed himself did not save him from being 'dragged through the mud' as he later wrote (*Werke*, VIII, p. 189).

However, if Gauss was the first person to think in this way, he was not the first person to vindicate doing so. He never proved there were no self-contradictions in a non-Euclidean geometry; he merely gave up looking for them and described a new geometry instead. Nor did he publish anything, afraid as he was of 'the clamour of the Boeotians' as he wrote (*Werke*, VIII, p. 200) in a letter to Bessel in 1829, and his work is known to us only in two unpublished manuscripts, some private notes, and in various letters to friends. I believe his timidity was due in main part to his knowledge that a decisive watertight construction of the new geometry was lacking. Since in many letters he confirms the work of his friends and sometimes extends it, it is hard to discover just how much he had found for himself, how much suspected,

---

[1] *Werke* (2nd edn), Vol. VIII, p. 177; this volume and Vol. IV contain much of Gauss's work on non-Euclidean geometry, together with excellent notes by P. Stäckel.

and how much was new. An odd characteristic of the mathematical mind is that it subconsciously re-orders material even when the problem has been put to one side, so that on returning to it new connections and consequences are at once apparent. Heuristic thinking of this kind is impossible to date, and the attribution of ideas is made even harder.

Gauss attended the University of Göttingen, where Kästner was Professor of Mathematics, from 1795 to 1798. A fellow student, Farkas Bolyai, was also interested in the problem of parallels, and he took his interest back to Hungary where in due course it infected his son Janos. At Göttingen in 1795 Gauss set about repairing the gaps in his mathematical education, and twice borrowed Lambert's *Theory of parallels* from the university library (Dunnington 1955).

Gauss's first work on parallels, written up in two memoranda, was based on classical methods. It appears to be predicated on the assumption that a new geometry is possible, but would accommodate itself to the discovery of a contradiction. The ambiguity is characteristic of work at this time and strikes deeper into mathematics than anyone at the time could have supposed. The ground is vanishing beneath their feet; they dare not look down. Yet the work is, quite literally, elementary. In the first memorandum, begun[2] it would seem in 1792 (when Gauss was 15 and before he ever went to Göttingen), Gauss discussed parallel lines.

The figure he had in mind is Saccheri's figure of asymptotic lines. His interpretation of it is new; for him it furnished a definition of parallel, which would seem to be essential to any attempt to build up a new geometry. Let us for now attempt to abstract from Euclid's geometry such properties of parallels as seem to be essential. We take the 1795 formulation of Playfair which was also known to Proclus. Given a line l and a point P not on it we can draw one and only one line through P which does not meet l. This line l' is called the parallel to l through P. If P' is also on l', then l' is the unique parallel to l through P, i.e. l' is everywhere parallel to l, and furthermore it follows from this that we can prove

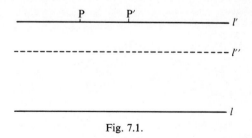

Fig. 7.1.

---

[2] Gauss wrote them down only in 1831, when, he said, they were 40 years old (*Werke*, VIII, pp. 202–8).

(1) *l* is parallel to itself
(2) if *l* is parallel to *l'* then *l'* is parallel to *l*
(3) if *l* is parallel to *l'* and *l'* is parallel to *l''* then *l* is parallel to *l''*.

These theorems are to be found in Euclid immediately after his definition of parallels, but it is amusing to prove them yourself. They are not difficult, although (1) is a little bit artificial, and they articulate the concept and make it easy to use. In order, they show that the property of being parallel is (1) reflexive, (2) symmetric and (3) transitive.

Such a property is frequently encountered in mathematics and is called an 'equivalence relation'. As the name implies it establishes a parity amongst the parallels to a given line: if it can be shown that one has a certain relevant property then, most likely, so do all the rest.

If you have established properties (1), (2), and (3) you will have noticed that the definition of parallels has two essential features. Firstly the existence (of *l'*) is guaranteed and secondly (for each point P) it is unique. Without either we do not obtain an equivalence relation and cannot begin any serious work. Gauss, we may imagine, reasoned similarly.[3] The problem is that the non-Euclidean geometries must abandon either the existence or the uniqueness to be non-Euclidean at all.

Let us return to Saccheri's figure (Fig. 4.12). In the pencil of lines through P two, *l'* and *l''*, are of interest, being asymptotic to *l*, and above them lies a family of lines not meeting *l* at all. (The real existence of the figure follows from assuming the HAA; it is not a misleading diagram.) Contrariwise, if we assume the existence of asymptotic lines like this we are led to the HAA. (The HOA we may take as destroyed.) So the figure contains numerous lines not meeting *l*, and if by parallel lines we mean those which do not meet we have existence but not uniqueness. Now draw only one half of the figure, say the lines extending rightwards. *l'* is the unique line asymptotic to *l*. In the sense of our earlier discussion it is the first line of the pencil through P which does not meet *l*. What Gauss did was to start here. He considered only *directed* lines, i.e. those extending in one direction from a specified point but which could be extended backward to another origin and thus indefinitely far backwards if so desired. He defined parallels as follows. Given a directed

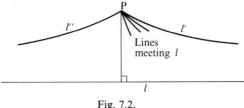

Fig. 7.2.

[3] The first substantial use of the concept of an equivalence relation is generally agreed to be found in Gauss's number-theoretical works of this period.

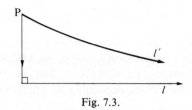

Fig. 7.3.

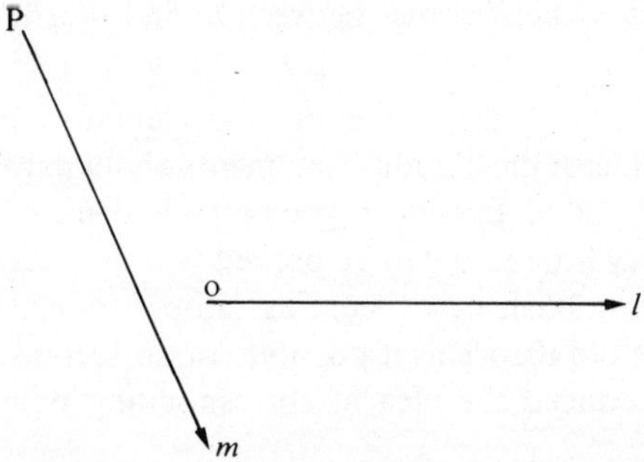

Fig. 7.4.

line l through O and a point P not on it the directed line parallel to l through P is the first directed line of the pencil through P which does not meet l: l' in Fig. 7.3. As the picture suggests, only directed lines in one direction are considered.

In Euclidean geometry there is only one such directed line, which must necessarily be the first. If Euclidean geometry is not the case then there will be several such lines, and the first is taken. We must of course allow the directed lines to be extended backwards indefinitely to avoid trivialities such as where, as drawn, m does not meet l. That we can do this is proved below.

On extending directed lines in Fig. 7.3 backward we, as it were, draw Saccheri's complete figure. Reversing the directions gives us the mirror figure, in which the line l has had its direction reversed and should properly be given a different name, because direction is important. The line l, undirected, inspires two directed lines through P which do not meet it (l' and l'' in the original figure) distinguished by their directions (right and left). To recapture uniqueness in the definition of parallels it is necessary to give each directed line its direction.

Gauss may now talk of unique parallel (directed) lines, so obviously the next task is to derive again the characteristic properties of parallels abstracted from the Euclidean case.

Gauss did this in an early memoir, showing that the new definition of parallels had the properties of being reflexive, symmetric, and transitive, and also that the apparent involvement of the point P in the definition is illusory (it really is the lines that are parallel, and if parallel at all they are parallel everywhere). The proofs are difficult exercises in conventional geometrical methods.

Exercises

7.1 Given directed lines l and m, show that if l is parallel to m then m is parallel to l. Hint: Take a point A on l, and draw AB perpendicular to m at B, AN any line making an angle α with l at A, and AC making an angle $\alpha/2$ with AB at A. There are two cases, depending on whether or not AC meets m.

7.2 Given three directed lines, l parallel to m and m parallel to n, show that l is parallel to n. Hint: either l comes between m and n or it does not.

With these theorems Gauss made the first tentative entry into a new geometry in which at least the obvious statements about parallels can be made. However, the vast bulk of Euclidean geometry is denied to us by our new assumption since very little in it can be proved without reference to parallels. The next task is to establish new theorems, proper to a non-Euclidean geometry, guided by the old theorems if possible. In his second memorandum on parallels Gauss introduced the idea of corresponding points on a pencil of lines.

In the case of Euclidean geometry there are two pencils: (i) the pencil of lines through a point and (ii) the pencil of lines perpendicular to a fixed line, shown vertical. In the case of Euclidean geometry this is also the pencil of parallels. In non-Euclidean geometry they must be distinguished; lines perpendicular to a fixed line are no longer parallel to each other but are 'ultra-parallel'.

For each type of pencil we take a point on one of them and locate the points on each line which correspond to it. Points A and B *correspond* with respect to a pencil if A lies on l (say) and B on m and AB makes equal angles with l and m. Such a set of points will be a locus. In Euclidean geometry the loci are as follows: for the pencil of lines through a fixed point the locus is a circle, and for the pencil of lines with a common perpendicular, the locus is another common perpendicular. In non-Euclidean geometry the same two loci exist as for Euclidean geometry plus, for the pencil of parallel lines, an unusual locus subsequently called the horocycle by Lobachevskii. The horocycle is not a circle and indeed has the remarkable property that no three points on it can be joined by a circle (see pp. 100–1). The horocycle plays an important role in non-Euclidean geometry, but Gauss did not develop any of its properties in this early memoir.

7.3 The construction for the circle through three given, non-collinear, points ABC in Euclidean geometry is as follows: draw the perpendicular bisectors p, q, r of BC, CA, AB respectively. They meet in a point O (why?) which is the centre of the required circle because O lies on p and is therefore equidistant from B and C, etc. Show that in non-Euclidean geometry it might happen that p, q, r are parallel or ultra parallel, in which case there is no circumcircle (a result first obtained by F. Bolyai).

Gauss's contribution 81

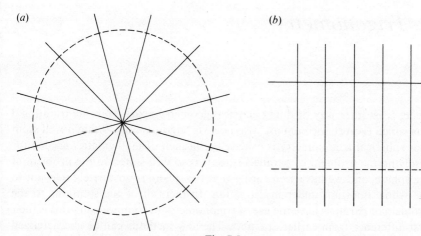

Fig. 7.5.

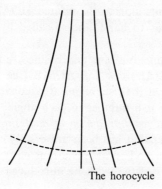

The horocycle

Fig. 7.6.

8 Trigonometry

There is another way of doing classical geometry apart from the traditional one which is more use in many ways, i.e. via trigonometry. It might well seem that spherical trigonometry is exactly the branch of mathematics needed for resolving the problem of parallels since it is so well suited to the problem of the sphere, and we have seen earlier that the sphere is obstinately relevant to the whole question under investigation. In fact, all of the solutions to the problem of parallels have the use of trigonometry in common, and this is their first difference from earlier attempts. The new methods can be characterized as analytic because of the use of the trigonometric functions to be described below. Of course, the usual trigonometric functions cannot be precisely the ones required, because in the plane they yield Euclidean results (HRA) and on the sphere results akin to the refuted HOA. By 1760, however, the right generalization was to hand: the so-called hyperbolic (trigonometric) functions.

The trigonometric functions sine and cosine are usually first encountered as ratios: $\sin A = BC/AB$; $\cos A = AC/AB$ (see Fig. 8.1). However, for purposes more advanced than surveying it is convenient to regard them as functions associating a real number to each angle. The way to do this was laid down by Euler in his *Introductio in analysin infinitorum* (1745, 1748, Lausanne) in a manner that has been universally copied since; indeed it is from this work that the modern abbreviations derive—Euler wrote sin., cos., and tang.. In Chapter 7 of the work he introduced the number e defined as the sum of the series

$$1 + 1 + \frac{1}{2!} + \frac{1}{3!} + \cdots$$

where $3! = 3 \times 2 \times 1$ and $n! = n(n-1)\ldots 2 \times 1$ and defined e^x to be the series

$$1 + x + \frac{x^2}{2!} + \frac{x^2}{3!} + \cdots$$

where x may be real, imaginary, or complex. He then turned to the trigono-

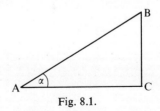

Fig. 8.1.

metric functions, defined as ratios, and sought power series expressions for them.

Since $\sin^2 x + \cos^2 x = 1$, by Pythagoras' Theorem, he factorized, obtaining

$$(\cos x + i \sin x)(\cos x - i \sin x) = 1$$

remarking 'the factors, though imaginary, are still of great use in combining and multiplying sines and cosines'. He then proved de Moivre's formula

$$(\cos x + i \sin x)^n = \cos nx + i \sin nx$$

from which he obtained a formula for $\cos nx$ as a polynomial in $\cos x$ and $\sin x$ which he was able to turn into a formula for $\cos x$ in terms of x, namely

$$\cos x = 1 - \frac{x^2}{2!} + \frac{x^4}{4!} - \cdots$$

and a formula for $\sin x$

$$\sin x = x - \frac{x^3}{3!} + \frac{x^5}{5!} - \cdots$$

He finally derived

$$\cos x = \frac{e^{ix} + e^{-ix}}{2}$$

and

$$\sin x = \frac{e^{ix} - e^{-ix}}{2i}$$

To obtain the power series for $\cos x$ Euler first let x be infinitely small, as he put it, so that $\sin x = x$ and $\cos x = 1$, and then let n be infinitely large while nx was finite, $nx = v$ say. Then, since $\sin x = x = v/n$,

$$\cos v = 1 - \frac{v^2}{2!} + \frac{v^4}{4!} - \cdots.$$

We can take different routes through these formulae. For example, we can define

$$\cos x = \frac{e^{ix} + e^{-ix}}{2}$$

$$i \sin x = \frac{e^{ix} - e^{-ix}}{2}$$

and derive de Moivre's formula by observing that

$$\cos nx + i \sin nx = e^{inx}$$
$$= (e^{ix})^n$$
$$= (\cos x + i \sin x)^n$$

as was to be proved.

Now the expressions

$$\frac{e^x + e^{-x}}{2} \quad \text{and} \quad \frac{e^x - e^{-x}}{2}$$

are obviously interesting. They are given the names cosh and sinh (pronounced to taste, usually 'cosh' or hyperbolic cosine, and 'shine' or hyperbolic sine) and called the hyperbolic functions:

$$\cosh x = \frac{e^x + e^{-x}}{2}$$

$$\sinh x = \frac{e^x - e^{-x}}{2}$$

and of course $e^x = \cosh x + \sinh x$. Their behaviour mimics that of their trigonometric cousins, except that $\sin \times \sin$ must always be replaced by $-\sinh \times \sinh$ (note the minus sign).

It is this parallel, with this strange twist, that is exploited to make the breakthrough. Once a formula is known in trigonometry, its analogue in hyperbolic functions can be written down and it is straightforward to prove it. This resulted in considerable labour saving since the formulae of trigonometry were well known to every contemporary mathematician. The exercises at the end of the chapter are devoted to this transition.

Interestingly, Johann Lambert was one of the first to introduce the hyperbolic functions as infinite series and to prove that they were indeed real valued. However, he does not seem to have worked with them in non-Euclidean geometry.

The best explanation of the name hyperbolic functions was also given by Lambert in his paper *Observations trigonometriques* (1768) where he gave them those names. The reason is, of course, the connection with the rectangular hyperbola. In the Fig. 8.2, taken from this paper, Lambert called Cp the hyperbolic cosine and pq the hyperbolic sine, whereas CM and MN are the circular cosine and sine. He then made great use of the formulae for cosh and sinh in terms of e and the strikingly similar expressions for cos and sin to obtain

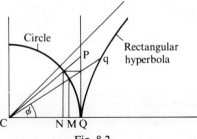

Fig. 8.2.

formulae in hyperbolic trigonometry analogous to the ones in ordinary trigonometry and to explain the correspondence. He even had a use for the hyperbolic functions in trigonometric problems: he applied them to problems in spherical trigonometry, and particularly astronomy on occasions when one of the celestial bodies is below the horizon. These problems can be solved with the familiar formulae of spherical trigonometry provided the arcs are taken to be pure imaginary. Not only does this motivate the transition from circular to hyperbolic functions, but it provides a plausible example of what an imaginary sphere might have meant to Lambert: one in which the angles are pure real and the sides of all triangles upon it pure imaginary. This has been pointed out by several writers (e.g. Peters 1961, Manning 1975) although Lambert nowhere says so explicitly.

Exercises

8.1 Prove
$$\cos^2 x + \sin^2 x = 1$$

I shall give two answers to this one, and then let you loose on the rest:

(a) $\cos x = \dfrac{e^{ix} + e^{-ix}}{2}$

Therefore
$$\cos^2 x = \frac{e^{2ix} + 2 + e^{-2ix}}{4}$$

$$\sin x = \frac{e^{ix} - e^{-ix}}{2}$$

therefore
$$\sin^2 x = \frac{e^{2ix} - 2 + e^{-2ix}}{-4}$$

therefore
$$\cos^2 x + \sin^2 x = \frac{e^{2ix} + 2 + e^{-2ix} - e^{2ix} + 2 - e^{-2ix}}{4}$$
$$= 4/4 = 1$$

as was to be proved;

(b) $\cos^2 x + \sin^2 x = (\cos x + i \sin x)(\cos x - i \sin x)$
$$= e^{ix} e^{-ix}$$
$$= e^0 = 1$$

as was to be proved.

8.2 Prove $e^{-ix} = \cos x - i \sin x$.
Hint: What are $\cos(-x)$ and $\sin(-x)$?

8.3 Prove $\cosh^2 x - \sinh^2 x = 1$.

8.4 Prove $\sin 2x = 2 \sin x \cos x$.

8.5 Prove $\cos 2x = \cos^2 x - \sin^2 x$.

8.6 Calculate $\sin 3x$ and $\cos 3x$ using de Moivre's formula. Hint:

$$e^{3ix} = \cos 3x + i \sin 3x$$

by writing $3x$ for x but

$$e^{3ix} = (\cos x + i \sin x)^3$$

by cubing both sides. Use $a + bi = c + di$ implies $a = c$ and $b = d$.

8.7 Calculate $\sinh 2x$ and $\cosh 2x$.
Check your answers against Exercises 8.4 and 8.5 using the analogue.

8.8 Calculate $\sin(x+y)$ and $\cos(x+y)$. Your answers should reduce to those to Exercises 8.4 and 8.5 in the special case of $x = y$.

8.9 Calculate $\sinh(x+y)$ and $\cosh(x+y)$. Check them (how?).

8.10 Show $\cosh x \geq 1$. Hint: Use Exercise 8.3 or the definition. N.B. For a fixed a and b, positive, the points in the plane $(x, y) = (a \cos \theta, b \sin \theta)$ define an ellipse, whose equation is

$$\frac{x^2}{a^2} + \frac{y^2}{b^2} = 1$$

Use Exercise 8.3 to show that the points $(x, y) = (a \cosh \theta, b \sinh \theta)$ lie on the curve

$$\frac{x^2}{a^2} - \frac{y^2}{b^2} = 1.$$

This curve is a hyperbola, whence the name hyperbolic functions.

8.11 Show $\sin(ix) = i \sinh x$ and $\cos(ix) = \cosh x$. This justifies the analogy between trigonometric and hyperbolic formulae.

9 The first new geometries

Gauss's correspondence with others, chiefly astronomers, on the emerging non-Euclidean geometry has already been mentioned. The two men who first moved the investigations into the area of analysis were also correspondents of his: F. K. Schweikart, Professor of Jurisprudence at Marburg, and his nephew, F. A. Taurinus.

Schweikart had read Lambert's work, as he mentions it in his book[1] of 1807, but he also established new results, most notably the one he sent to Gauss in a letter of 1818 which concerned what he called an Astral Geometry (*Werke*, VIII, p. 180):

... That the altitude of an isosceles right-angled triangle continually grows, as the sides increase, but can never be greater than a certain length, which I call the *constant*.

'Squares have therefore the following form.'[2] (Quoted by Bonola, p. 76) (Fig. 9.1(a)).

Schweikart sent his note to Gauss through the intermediary of Gerling, a mathematician at Marburg. On 16 March 1819 Gauss replied to Gerling (*Werke*, VIII, p. 181):

The letter of Herr Professor Schweikart has given me extraordinary pleasure, and I would really like to say a lot of good things to him about the work ... I shall only note that I can solve completely all the problems in the Astral Geometry—so far as it has been developed—as soon as the constant C is given.

In particular he observed that the area of a triangle is proportional to its angular defect (2π − angle sum) and that an asymptotic triangle, one in which the three pairs of sides are asymptotic to each other, has area $\pi C^2 / \{\log(1 + \sqrt{2})\}^2$. Schweikart's nephew Taurinus became interested and

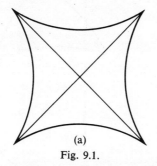

Fig. 9.1.

---

[1] *Theorie der Parallellinien*, an orthodox treatment (see Bonola, p. 77).
[2] Squares were also drawn like this by Levi ben Gerson in the fourteenth century, who also remarked that now the diagonal can be commensurable with the side (see Toth 1969).

88 The first new geometries

published two books in 1825 and 1826. He was able to relate the constant to other quantities occurring in the geometry, and yet he continued to believe that no new geometry was possible for space, arguing from the absolute nature of length to conclusions which violated his conception of space. This is an example of how determined people are to prove the conclusions that they wish to hold. Taurinus's two works are quite different in character. The 1825 publication is hostile to non-Euclidean geometry, which is 'repugnant to all intuition', and several objections are made against it. Unhappily, they are all intellectually shallow and reflect the heavy strain imposed on anyone who tries to make things new. By contrast in 1826 we find him writing of a 'log-spherical geometry' which, he conjectures, will be a geometry of some sort but not possible on a plane.

Taurinus's method was to transcribe the formulae of spherical trigonometry in the following way.

The first formula connecting the sides and angles of a spherical triangle is

$$\cos \frac{\alpha}{K} = \cos \frac{\beta}{K} \cos \frac{\gamma}{K} + \sin \frac{\beta}{K} \sin \frac{\gamma}{K} \cos \hat{A}$$

where K is the radius and α, β, γ, and A are the angles shown (Fig. 9.1(b)). Replacing K by ik is sensible algebraically (although it is a doubtful geometric manoeuvre giving birth to the 'imaginary sphere') because $\cos ik$ and $i \sin ik$ are still real. The formula becomes

$$\cosh \frac{\alpha}{K} = \cosh \frac{\beta}{K} \cosh \frac{\gamma}{K} - \sinh \frac{\beta}{K} \sinh \frac{\gamma}{K} \cos \hat{A} \tag{9.1}$$

the minus sign arising for the reason discussed above.

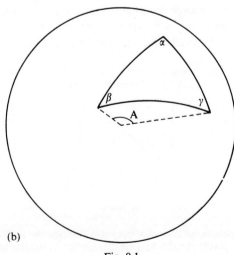

(b)

Fig. 9.1.

We notice straight away that the angles of a triangle must be less than 180°; taking for simplicity an equilateral triangle where $\alpha = \beta = \gamma$ we obtain

$$\cosh \frac{\alpha}{K} = \cosh^2 \frac{\alpha}{K} - \sinh^2 \frac{\alpha}{K} \cos \hat{A}.$$

Therefore

$$\cos A = \frac{\cosh^2 (\alpha/K) - \cosh (\alpha/K)}{\sinh^2 (\alpha/K)}$$

$$= \frac{\{\cosh (\alpha/K) - 1\} \cosh (\alpha/K)}{\cosh^2 (\alpha/K) - 1} \qquad \text{(by Exercise 8.3)}$$

$$= \frac{\cosh (\alpha/K)}{\cosh (\alpha/K) + 1}$$

but $\cosh (\alpha/K) > 1$ since $\alpha \neq 0$, and therefore

$$\cos \hat{A} > \tfrac{1}{2}$$

and

$$\hat{A} < 60°$$

Therefore the angle sum is less than 180°.

Furthermore, as the sides become smaller, the angle become larger; precisely

$$\lim_{\alpha \to 0} (\cos \hat{A}) = \tfrac{1}{2} \quad \text{since} \quad \lim_{\alpha \to 0} \cosh (\alpha/K) = 1.$$

Therefore the angle sum tends to 180° ($\cos 60° = \tfrac{1}{2}$) and small triangles differ very little from Euclidean ones.

We can also show that as K becomes larger the same thing happens.

$$\lim_{K \to \infty} \cosh (\alpha/K) = \cosh (0) = 1$$

and so

$$\lim_{K \to \infty} \frac{\cosh (\alpha/K)}{\cosh (\alpha/K) + 1} = \frac{1}{2}.$$

Therefore as K becomes larger and larger the more nearly all triangles on it become Euclidean.

Indeed, in the limit the formula (9.1) reduces to

$$\alpha^2 = \beta^2 + \gamma^2 - 2\beta\gamma \cos \hat{A}$$

the fundamental formula of Euclidean plane trigonometry (the proof is a little cluttered and is given in the Appendix to this chapter).

90   *The first new geometries*

There is a second fundamental formula in spherical trigonometry:

$$\cos \hat{A} = -\cos \hat{B} \cos \hat{C} + \sin \hat{B} \sin \hat{C} \cos \frac{a}{K}$$

which becomes, with hyperbolic functions,

$$\cos \hat{B} = -\cos \hat{B} \cos \hat{C} + \sin \hat{B} \sin \hat{C} \cosh \frac{a}{K}$$

A special case of this is when $\hat{A} = 0$ and $\hat{C} = 90°$, when

$$1 = \sin \hat{B} \cosh \frac{a}{K}$$

or

$$\cosh \frac{a}{K} = \frac{1}{\sin \hat{B}}$$

The triangle described by this formula has a right angle at C and BA asymptotic to CA, so the vertex A is 'removed to infinity'. It is best thought of as a limiting case not actually obtained. The formula gives the relationship of β to a, and we have obtained a new and sharp description of the familiar figure of Saccheri and Gauss for the HAA. We shall henceforth call the angle at B the *angle of parallelism* for a, since the line through B making that angle with BC is parallel to CA (recall Proclus's remark, p. 39).

We may now, still following Taurinus, derive Schweikart's constant, i.e. the maximum possible altitude of an isosceles right-angled triangle (WZ in triangle WXY, where WX = WY), in terms of K. Note that $Z\hat{W}Y = Z\hat{W}X = \pi/4$, by symmetry.

It is most useful if we draw the figure as shown in Fig. 9.3, and it is now clear that △ YWX reaches maximum area when WX is asymptotic to ZX, for as W goes higher YZX becomes larger, but if W rises beyond the point at which WX is asymptotic to ZX there is no triangle to measure. This occurs when the angle of parallelism is 45° and WZ is then Schweikart's constant which

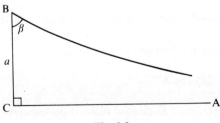

Fig. 9.2.

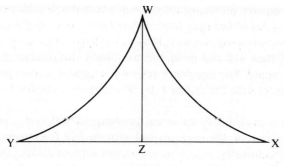

Fig. 9.3.

we shall call C. Drawing that figure (half is enough) we are confronted with the familiar Fig. 9.2. From the argument above, therefore,

$$\cosh \frac{C}{K} = \frac{1}{\sin 45°} = \sqrt{2}$$

which we can solve to obtain

$$K = \frac{C}{\log(1+\sqrt{2})}$$

(see the appendix to this chapter for the proof), thereby relating C and K.

Taurinus also obtained other results by the same means, concerning amongst other things the areas of triangles, the circumference of circles, and the area and volume of spheres. However, we are more concerned with the significance of his technique than with the results it obtained.

Taurinus himself commented in his book of 1826 (§ 64):

This [the book] was already written and it seemed to me to remain to state my views on the true essence of this Geometry. I am led at last to the certainty that my answer is really proved. From the very beginning I have conjectured that a geometry so to speak inverse to spherical (the logarithmic) with all formulae derivable from the spherical can exist. And I have further wondered that this fact, which is so clear and lies so readily to hand, was not spotted, and such a vast an extent not explored until I recalled that every self-evident matter is often hidden even for a long time even from the most perceptive of men. Moreover, I thought that in everything that was earlier deduced concerning the analytic formulae there was nothing that refers to this geometry, and that by pure substitution the formulae remain valid.

None the less he did find reason to adhere to his earlier (1825) objections to the new geometry as being incompatible with our notions of space, but unhappily his reasoning is shallow here.

His method is entirely algebraic; to talk of the sphere of imaginary radius makes very little, if any, sense, geometrically. So for the first time we encounter a method which is not pictorial, but merely formal. As equations, what he has written is correct, but what does it describe? You may well feel that it is an excursion, pleasant enough no doubt, but well aside from the main topic. The formulae of ordinary trigonometry apply to triangles since they are painstakingly constructed that way. If by some sleight of hand we can produce new formulae, then well and good, but we should also produce the geometry to which they apply. The imaginary sphere is a figment, a mere paraphrase of the instruction 'replace the radius k by ik wherever it occurs' but devoid of content.

The situation in ordinary spherical geometry is different, and may well serve as an illustration. We may proceed with the two fundamental formulae of spherical trigonometry, exactly as above, but without making the transition to hyperbolic functions. The conclusions we reach algebraically are of course similar to the above, except that we have that the angles of an equilateral (and hence any) triangle must sum to greater than 180°. We have

$$\cos \frac{\alpha}{K} = \cos \frac{\alpha}{K} \cos \frac{\alpha}{K} + \sin \frac{\alpha}{K} \sin \frac{\alpha}{K} \cos \hat{A}.$$

Therefore

$$\frac{\cos(\alpha/K)\{1 - \cos(\alpha/K)\}}{\sin^2(\alpha/K)} = \cos \hat{A}$$

and

$$\frac{\cos(\alpha/K)}{\cos(\alpha/K) + 1} = \cos \hat{A}$$

but now $\cos(\alpha/K) < 1$ and so

$$\cos \hat{A} < \tfrac{1}{2} \quad \text{and} \quad \hat{A} > 60°$$

We again find that small triangles are nearly Euclidean.[3]

Asymptotic lines, however, are not a feature of spherical trigonometry, and the argument concerning the special case of the second formula breaks down utterly. However, there are similar results for the area of triangles, the circumference of circles, and the area and volume of spheres.

However, the formulae refer to a model, a picture, which is the sphere. It was Lambert who first pointed out the connection between spherical geometry and the geometry based on the HOA, and he also suggested that the AA'd geometry would be that on the imaginary sphere. He was himself also engaged on research into the hyperbolic functions, and at least some of Taurinus's results were known to him, but he never considered that they might describe a non-Euclidean geometry and only his conjecture remains. Taurinus for his part seems to have wanted a better surface (a *real* surface) to which his formulae applied, but he never believed that they described a geometry on the plane. They were formulae, suggestively related to the HAA, but geometrically meaningless.

[3] Gauss also proved this result (*Werke*, Vol. VIII, p. 255, 'Die Sphärische und die Nicht-Euklidische Geometrie').

We can imagine the HAA being revealed as self-contradictory despite the formal validity of Taurinus's equations. Indeed it is easy to do so, since the comparatively transparent case of the HOA and spherical geometry is somehow resolved. How much more leeway do we have then when the curious imaginary sphere is invoked? If a contradiction were to pop up we might either decide that this new type of sphere was a nonsense or that it was a valid object of study but irrelevant to the problem at hand.

However, there is a 'contrariwise' which must have haunted the researchers and has perhaps occurred to the reader. Geometry has begun to seem very dubious with all this ultimately inconclusive stuff about straight lines. By contrast the trigonometry is clear and rigorous. Its conclusions, although novel, might inspire more faith than the traditional results whose grounding in intuition can lead us astray. Is not a new geometry established by the formulae? If not one, why not both? The less adventurous can settle for the imaginary spherical geometry; a more bold approach would be to accept spherical geometry as well. That leaves one troublesome problem: the refutations of the HOA supplied by Saccheri and Legendre amongst others; one would hope to learn more about the implicit assumptions of Euclidean geometry until one could refute the 'refutations'. Far from being the conclusive logical system geometry is beginning to seem ambiguous and obscure.

The less adventurous position was in fact the one adopted. It came to be thought that a plane geometry on which the HAA obtained was possible, but the refutations of the OA'd case stood unchallenged. It is with the formulation of these ideas that the two most famous names in non-Euclidean geometry are associated: Lobachevskii and Janos Bolyai.

Appendix

(1) We have

$$\cosh \frac{\alpha}{K} = \cosh \frac{\beta}{K} \cosh \frac{\gamma}{K} - \sinh \frac{\beta}{K} \sinh \frac{\gamma}{K} \cos \hat{A}$$

Therefore

$$\left(1 + \frac{\alpha^2}{K^2 2!} + \cdots\right) = \left(1 + \frac{\beta^2}{K^2 2!} + \cdots\right)\left(1 + \frac{\gamma^2}{K^2 2!} + \cdots\right)$$
$$- \frac{\beta}{K}\left(1 + \frac{\beta^2}{K^3 3!} + \cdots\right) \frac{\gamma}{K}\left(1 + \frac{\gamma^2}{K^2 3!} + \cdots\right) \cos \hat{A}$$

Therefore

$$\frac{\alpha^2}{2K^2} + \text{terms in } K^{-4} \text{ etc.} = \frac{\beta^2}{2K^2} + \frac{\gamma^2}{2K^2} + \text{terms in } K^{-4} \text{ etc.}$$
$$- \left(\frac{\beta\gamma}{K^2} + \text{terms in } K^{-4}\right) \cos \hat{A}.$$

94 The first new geometries

Multiplying by K^2

$$\frac{\alpha^2}{2} + \text{terms in } K^{-2} \text{ etc.} = \frac{\beta^2 + \gamma^2 - 2\beta\gamma \cos \hat{A}}{2} + \text{terms in } K^{-2} \text{ etc.}$$

As $K \to \infty$, $K^{-2} \to 0$ and the equality becomes

$$\alpha^2 = \beta^2 + \gamma^2 - 2\beta\gamma \cos \hat{A}$$

as promised.

(2) The proof is best done back to front.

$$K = \frac{C}{\log(1 + \sqrt{2})};$$

therefore

$$\log(1 + \sqrt{2}) = \frac{C}{K}$$

and

$$1 + \sqrt{2} = e^{C/K}. \tag{9.2}$$

Also

$$\log \frac{1}{1 + \sqrt{2}} = -\log(1 + \sqrt{2}) = -\frac{C}{K};$$

therefore

$$\frac{1}{1 + \sqrt{2}} = e^{-C/K} \tag{9.3}$$

but

$$\frac{1}{1 + \sqrt{2}} = \sqrt{2} - 1.$$

Therefore, adding (9.2) and (9.3)

$$1 + \sqrt{2} + \sqrt{2} - 1 = e^{C/K} + e^{-C/K} = 2 \cosh \frac{C}{K}$$

and

$$\sqrt{2} = \cosh \frac{C}{K}.$$

To prove our result we can run this argument backwards which we could have done anyway, but it would have looked a little devious.

Exercise

9.1 Prove the formulae for the area and circumference of a circle on a sphere where the radius of the circle (as measured by a geodesic on the surface) is r and the radius of the sphere is R:

$$\text{area circle} = 4\pi R^2 \sin^2 \frac{r}{2R}$$

$$\text{circumference circle} = 2\pi r \sin \frac{r}{R}.$$

10 The discoveries of Lobachevskii and Bolyai

The work of Bolyai and Lobachevskii is astonishingly similar, and yet each remained in ignorance of the very existence of the other until some years after their work was published.[1] The Hungarian was aware through his father of the Western European work on non-Euclidean geometry, although he appears to have largely gone his own way. The Russian kept apart from the controversy, and his earliest publication on the subject, in the *Kazan Messenger* for 1829, passed completely unnoticed.

Both began work on the parallel problem convinced that they would find the decisive refutation of the alternative hypotheses. It was the road to despair. Farkas Bolyai counselled his son:

You must not attempt this approach to parallels. I know this way to its very end. I have traversed this bottomless night, which extinguished all light and joy of my life. I entreat you, leave the science of parallels alone . . .
I turned back when I saw that no man can reach the bottom of this night. I turned back unconsoled, pitying myself and all mankind.

Elsewhere he wrote:

I admit that I expect little from the deviation of your lines. It seems to me that I have been in these regions; that I have travelled past all reefs of this infernal Dead Sea and have always come back with broken mast and torn sail. The ruin of my disposition and my fall date back to this time. I thoughtlessly risked my life and happiness—*aut Caeser aut nihil*. [Quoted by Meschkowski 1964].

Both gradually changed their minds. By 1823 Lobachevskii was willing to consider an imaginary geometry, and in 1826 he gave a lecture in the School of Mathematical Physics at Kazan University in which a geometry was outlined in which through a point there are two lines parallel to a given one. Plainly this is the AA'd geometry. Unhappily the manuscript to this lecture is lost, but a reworking of it is contained in the *Memoir* of 1829. Successive publications followed,[2] including a summary of his work, *Geometrical researches on the theory of parallels*, which was written in German in 1840 and was intended to reach a European audience. We shall take our account of his work from this publication. Halsted's translation of it is printed as an Appendix in Bonola's book along with his translation of Bolyai's paper and some useful historical notes.

---

[1] Janos Bolyai did not read Lobachevskii's papers until 1848; his father compared the two treatments in his *Kurzer Grundriss* of 1851.

[2] Lobachevskii's two earlier works are more or less recapitulated in his German paper of 1840. They were translated into German themselves in 1899. The chief difference between them and the 1840 paper is the greater attention paid in the earlier papers to the ideas of space, contiguity, and distance.

The discoveries of Lobachevskii and Bolyai   97

Bolyai was convinced until 1821 that the parallel postulate must hold, and the tenacity of his mistaken belief perhaps explains the force with which he subsequently came to explore the opposite point of view. As early as 1823 he believed that he was near success, and wrote the following marvellous letter to his father:

I have now resolved to publish a work on parallels ... I have not yet completed the work, but the road that I have followed has made it almost certain that the goal will be attained, if that is at all possible. the goal is not yet reached, but I have made such wonderful discoveries that I have been almost overwhelmed by them, and it would be the cause of continual regret if they were lost. When you see them, you too will recognize them. In the meantime I can say only this: *I have created a new universe from nothing*. All that I have sent you till now is but a house of cards compared to a tower. I am as fully persuaded that it will bring me honour, as if I had already completed the discovery. (3 November 1823).

We may imagine the excitement with which his father received the news. He replied:

... if you have really succeeded in the question, it is right that no time be lost in making it public, for two reasons: first, because ideas pass easily from one to another, who can anticipate its publication; and secondly, there is some truth in this, that many things have an epoch, in which they are found in several places, just as violets appear on every side in the Spring. Also every scientific struggle is just a serious war, in which I cannot say when peace will arrive. Thus we ought to conquer when we are able, since the advantage is always to the first comer.

In 1825 Janos had an abstract of his work ready and sent it to his father and his old professor, amongst others. In 1829, despite niggling doubts about it, the Bolyais agreed to publish it as an appendix to the father's book, the *Tentamen*. The book came out in 1831 and a copy was sent to Gauss, but it never arrived. A second copy was dispatched, reaching Göttingen in 1832. Gauss replied seven weeks later (6 March 1832):

If I commenced by saying that I am unable to praise this work (by Janos), you would certainly be surprised for a moment. But I cannot say otherwise. To praise it, would be to praise myself. Indeed the whole contents of the work, the path taken by your son, the results to which he is led, coincide almost entirely with my meditations, which have occupied my mind partly for the last thirty or thirty-five years. So I remained quite stupefied. So far as my own work is concerned, of which up till now I have put little on paper, my attention was not to let it be published during my lifetime. Indeed the majority of people have not clear ideas upon the questions of which we are speaking, and I have found very few people who could regard with any special interest what I communicated to them on this subject. To be able to take such an interest it is first of all necessary to have devoted careful thought to the real nature of what is wanted and upon this matter almost all are most uncertain. On the other hand it was my idea to write down all this later so that at least it should not perish with me. It is therefore a pleasant surprise for me that I am spared this trouble, and I am very glad that it is just the son of my old friend, who takes the precedence of me in such a remarkable manner. (*Werke*, VIII, p. 221; quoted by Bonola, p. 100.)

Farkas was satisfied with the reply, but Janos was not and never really forgave the Prince of Geometers. He was incredulous that Gauss should have made his discoveries before him, and became so convinced of plagiarism that he never published again.

Strangely, the works passed straight into obscurity, and it was Gerling (through whom Schweikart had earlier corresponded with Gauss) who decided to rescue them. Gauss's own comments and letters were published between 1860 and 1863 in the first edition of his *Werke*. Baltzer's *Elemente der Mathematik* (2nd edn, 1867) discussed geometry from the new point of view, giving prominence to the names of its inventors (Bonola, §62). Baltzer also persuaded Houel to translate the originals into French, and their publication in 1866 and 1867 came, as we shall see, at the right time. Houel remained a generous propagandist for the non-Euclidean geometry, and the explosion of translations and memoirs in various languages over the next ten years is largely due to him (the English translations are mostly by G. B. Halsted).

Let us now turn to the geometry of Bolyai and Lobachevskii.

Absolute geometry

A theorem will be said to be *absolute*, or a theorem in *absolute geometry*, if it holds independently of the parallel postulate (the word is Bolyai's). The trivial ones are the first 28 propositions of Euclid.

We shall encounter more important absolute theorems later on. Such a theorem might be obtained either without (even concealed!) reference to parallels, or else separately on assuming first the postulate and secondly the HAA. If the definition of parallel was so phrased as to make sense in either

Table 10.1. *A summary of Lobachevskii's geometrical researches on the theory of parallels as an aid to Chapter 10*

---

(1) Parallels are defined, in a space of two of three dimensions (p. 99).

(2) A pencil of lines in space parallel to a given line is defined (p. 99).

(3) Numerous theorems concerning parallels and angle sums of triangles are proved, with which we are already familiar; these theorems are independent of the definition of parallels, and are called absolute (p. 100).

(4) A new theorem is proved: if three planes cut each other in three parallel lines then the angles between the planes sum to two right angles. We shall call this theorem the prism theorem (see p. 100, although the proof is omitted). This theorem is also an absolute theorem.

(5) An important curve is introduced, called the horocycle or boundary line. Its analogue in space is a surface called the horosphere (pp. 101, 101). Analytic formulae for the horocycle are introduced.

(6) A natural geometry is introduced onto the horosphere (p. 101) and a natural map from it to the hyperbolic plane is demonstrated (p. 101).

(7) The geometry on the horosphere is shown to be Euclidean because of the prism theorem (p. 102).

(8) The formulae of (3) and (5), together with the mapping (6) from horosphere to hyperbolic plane establish formulae in hyperbolic geometry for the hyperbolic plane (p. 106).

(9) These formulae are shown to resemble the usual Euclidean formulae in many ways (p. 107).

geometry, then the theorems we would obtain would be absolute. If, however, in the course of the work a special case was taken, one might then obtain theorems true only in Euclidean geometry, or only in non-Euclidean geometry, and if it further turned out that a special case was forced on us we might find a contradiction which would destroy the non-Euclidean geometry.

Lobachevskii's theory of parallels (1840).

Lobachevskii's definition of parallel (*Theory of parallels*, §16, to be referred to as TP) has just this ambiguous character. It is, predictably, as follows. l' is the parallel to l through P (rightwards) if it is the first line of the pencil of lines through P not to meet l. The angle $l'p$ at P, α, is called the angle of parallelism of l' and depends on the length of p. We shall write $P(p) = \alpha$.

The Euclidean case always corresponds to $\alpha = 90°$; the non-Euclidean case always corresponds to $\alpha < 90°$. Accordingly, if in a proof we say $\alpha = 90°$ our theorem is one of Euclidean geometry, and if we say $\alpha < 90°$ it is one in non-Euclidean. If we never specify, or it turns out not to matter because we can do it either way, our theorem is absolute. If, however, at any time the assumption $\alpha = 90°$ is forced on us because having $\alpha < 90°$ would lead to a contradiction, then Euclidean geometry would stand alone as the only possible geometry.

Lobachevskii proved the essential preliminary theorems, with the extra twist that he was working in three dimensions and so pencils of parallel lines now look like a prickly ball (see Fig. 10.3). He then turned to triangles drawn on spheres, and the 'solid angle' which they subtend at the centre. Solid angles are measured by comparing them with the total angle at a point, which is proportional to the area of the sphere, and so the total solid angle is 4π (compare the circle case, p. 65). If the solid angle is θ and the vertical angles are α, β, and γ, then $\theta = (\alpha+\beta+\gamma)-\pi$ (TP27). This result essentially expresses the connection between the area of a spherical triangle and its angular excess p that we proved earlier, but let us check it for a special triangle formed by the equator, the 0° longitude and the 90° longitude. Here $\alpha = \beta = \gamma = \pi/2$ and θ is one-eighth of the total angle at the centre and is therefore

$$\theta = \frac{4\pi}{8} = \frac{\pi}{2} = \left(\frac{\pi}{2}+\frac{\pi}{2}+\frac{\pi}{2}\right) - \pi$$

as required. (To agree with modern usage our units are twice those used by Lobachevskii.) He then proved the prism theorem.

The prism theorem

If three planes cut each other in parallel lines then the sum of the three surface angles is $\pi = 2R$ (TP28). The surface angle between two planes is defined to be the angle between the perpendiculars to the planes. Let the planes meet in the parallel lines a, b, and c, where the planes themselves form the walls of a 'prism'.

100 The discoveries of Lobachevskii and Bolyai

In the sequel it is necessary to know that a surface angle between two planes is cut out on any plane which meets the given planes at right angles. Therefore, to find the surface angle between planes P_1 and P_2 which meet along a line l it is enough to cut P_1 and P_2 with a third plane P_3 perpendicular to l. P_1 and P_2 cut P_3 in l_1, l_2 respectively, and the angle between l_1 and l_2 is equal to the surface angle between P_1 and P_2.

The lines a, b, and c are supposed to be parallel. It is the angles made by the walls which sum to $2R$, and notice this is an absolute theorem. Now the story really starts.

In the usual situation the three perpendicular bisectors of the sides of triangle meet in a point, the centre of the circumcircle (see p. 80). In the enlarged situation there is an alternative, namely that the three bisectors are all mutually parallel; once two of them are parallel the third one must be (TP30).

What of the curve drawn such that the triangle formed by any three points upon it has the property that the perpendicular bisectors are mutually parallel? It is the curve that we met in Gauss's work on corresponding points, which Lobachevskii calls the *horocycle* or *boundary line*.

His construction for it is this (TP31). Let A be a point and BA a line through A. A point C on the curve makes an angle of $\alpha = P(AC/2)$ with BA, where P is the angle of parallelism function, defined above (p. 90). By letting the centre of the circumcircle move away to infinity (i.e. by flattening the

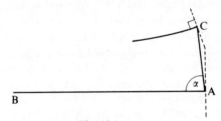

Fig. 10.1.

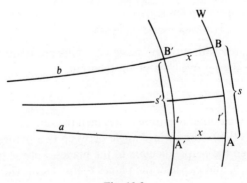

Fig. 10.2.

triangle) we see that a circle of continually increasing radius merges into the boundary line (TP32).

BA is perpendicular to the horocycle just constructed, and all of the perpendiculars to the horocycle are mutually parallel. We now observe the precise manner in which parallels are asymptotic (TP33). Let a and b be two perpendiculars to the horocycle w, meeting it at A and B respectively. Take A' on a and through it draw the horocycle to a. Suppose it meets b at B'.[3] We have $AA' = BB' = x$, say.

Let the lengths of the segments along the arcs be $AB = s$ and $A'B' = s'$. Then $s' = se^{-x}$. To prove this assume that the ratio $s/s' = n/m$, which we may by continuity, and take a third axis CC' to the horocycle. Let the arc lengths AC and A'C' be t and t' respectively, and suppose that $t/s = p/q$. Divide the arc s by axes into nq equal parts, $s = nq$ say; there will then be mq such parts on s and np on t, but by symmetry these axes divide s' and t' similarly, so $t'/t = s'/s$, i.e. as long as x does not change the ratio of t'/t is independent of t. Accordingly, if $x = 1$, $es' = s$ and generally $s' = se^{-x}$, where e could be any constant. We shall always take it to be the exponential e. As $x \to \infty$, $s' \to 0$, so we see that parallel lines are indeed asymptotic.

Of greatest important in Lobachevskii's work is the horosphere, which is obtained by rotating the horocycle about any one of its axes. (Rotating a circle in this fashion about a radius generates a sphere.) Since, by symmetry, the chord joining two points A and B on a horocycle is inclined equally to the axes at A and B, the same horosphere is obtained whichever axis is chosen (TP34). Any plane in space which contains an axis of the horosphere cuts it in a horocycle (TP34). The horosphere is also the locus of points corresponding to a given point with respect to a three-dimensional pencil of parallel lines, as can be seen by taking any line of the pencil and any plane containing that line and rotating the plane about that line as axis. In what follows it is important to remember that a unique pencil of parallel lines is associated with a horosphere.

Imagine now that you have a horosphere sitting upon a table. It rests upon a point A, and the axis through A is perpendicular to it and the table. The other axes to it fan out as shown in Fig. 10.3. Some of them also meet the table; some do not. Those which are asymptotic to the table meet the horosphere in a circle Ω, which does not seem to have interested Lobachevskii but which attracted attention later on. That there should be any such asymptotes can be seen by laying the picture on its side, when it becomes much more familiar.

By following down the axes we obtain a map from the horosphere to the plane (the table). This map we call the projection map. Under it horocycles go into straight lines. Accordingly, if we call a triangle on the horosphere that figure which is bounded by three horocycles, that is the intersections of three planes with the horosphere, we may investigate the geometry on the horosphere using the projection map.

---

[3] It must meet b. Can you use the earlier construction to see why?

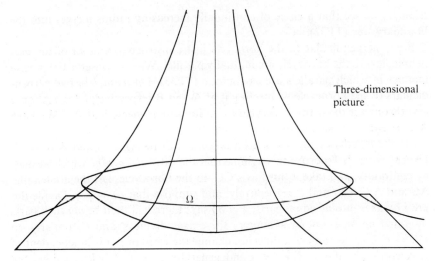

Fig. 10.3.

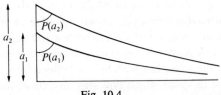

Fig. 10.4.

What is the angle sum of a triangle on the horosphere? Since the lines of the pencil each meet the horosphere at right angles and the plane formed by any two lines of the pencil meets the horosphere in a horocycle, the angle between two horocycles equals the angle between the two corresponding planes. Accordingly such a triangle is formed by three planes in the form of a prism, and we have already seen that the sum of the angles between the walls of a prism sum to $2R$. The geometry on the horosphere is Euclidean!

Lobachevskii next soared between Euclidean, spherical, and the new geometries in the following way (TP35) (Fig. 10.5).

$A''A$, $B''B$, and $C''C$ are three axes of a horosphere (not shown); A, B′, and C′ lie on the horosphere and A, B, and C lie on the plane. The triangle ABC has a right angle at C. The lengths of the sides of the triangle are $AB = c$, $BC = a$, and $CA = b$. The angles are $B\hat{A}C = P(\alpha)$ and $A\hat{B}C = P(\beta)$, say, for some α and β.

Draw a sphere of unit radius with centre B. A spherical triangle is obtained upon it called $A'''B'''C'''$, with sides p, q, and r, where $p = B'''C'''$, $q = C'''A'''$, and $r = A'''B'''$. We shall now determine the sides and angles in this spherical triangle. The sides are equal to the angles they subtend at B, the centre of the sphere. p subtends $B'''\hat{B}C''' = B'\hat{B}C$, but B′B and C′C are parallel, so $B'\hat{B}C = P(a)$, which is by definition of the angle of parallelism. Similarly, $r = P(c)$. q subtends the angle $A'''\hat{B}C'''$ which we have called $P(\beta)$, so we conclude that $q = P(\beta)$. To summarize

$$p = P(a), \quad q = P(\beta), \quad r = P(c).$$

We next find the angles in the spherical triangle. The angle $\hat{A}'''$ may be considered as the angle between the tangent to $A'''B'''$ and the tangent to $A'''C'''$. These tangents lie perpendicularly to the radius BA''', so they lie in a plane perpendicular to BA''', which is the line of intersection of the

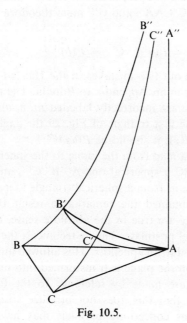

Fig. 10.5.

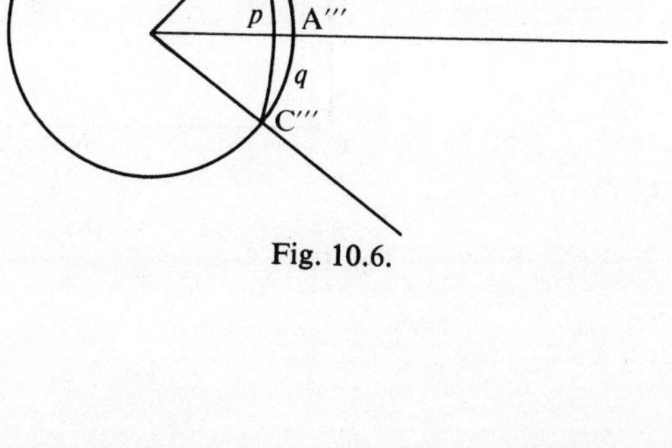

Fig. 10.6.

planes B″BAA″ and CBA. By the remark following the prism theorem it follows that the angle between those planes is the angle between the tangents. However, the angle between those planes is a right angle, since A″A is perpendicular to ABC: $\hat{A}''' = \pi/2$. In the same way the angle at B‴ is the angle between the planes B″BAA″ and B″BCC″, which is also the angle $A\hat{B}'C'$ on the horosphere (because the planes cut the horosphere at right angles) so

$$\hat{B}''' = \frac{\pi}{2} - P(\alpha) = P(\alpha')$$

say. Finally $\hat{C}'''$ lies in a plane perpendicular to BC‴ and so, since $B\hat{C}A = \pi/2$, in a plane parallel to C″CAA″, and $\hat{C}'''$ must therefore be equal to $C''\hat{C}A = P(b)$. To summarize

$$\hat{A}''' = \pi/2, \quad \hat{B}''' = P(\alpha'), \quad \hat{C}''' = P(b)$$

(I should now point out two mistakes in the Halsted translation of Lobachevskii's paper given in an Appendix to Bonola. The points of intersection of BA, BB‴, and BC are incorrectly labelled m, n, and k; they should be labelled n, m, k. In the first triangle of Fig. 29 the angle between b and c is incorrectly given as $P(a)$; it should be $P(\alpha)$.)

Therefore we have a map from the plane to the sphere which associates to a (linear) triangle ABC a spherical one A‴B‴C‴, and the process can be run backwards to take us from a spherical triangle to a specifiable linear one. Lobachevskii then obtained the remarkable result that the formulae of spherical trigonometry are true in the absolute sense, that is they are independent of the parallel postulate. This is technical; the simpler proof found by J. Bolyai is relegated to an appendix. This allowed him to use the following procedure. A figure on the plane is transformed into one on the sphere, and deductions about it are made by reference to the formulae of spherical trigonometry. Notice that this does not produce absolute theorems, since particular assumptions concerning parallels may have been made before passing over to the sphere. The fundamental formulae for a right-angled spherical triangle now can be interpreted as relations concerning a right-

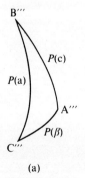
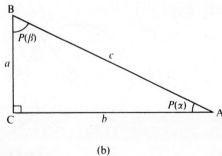

Fig. 10.7.

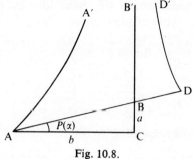

Fig. 10.8.

angled non-Euclidean triangle. They can further be used to obtain a description of the function $P(x)$ itself.

Lobachevskii considered angles $P(x-y)$ and $P(x+y)$ for $y \leq x$, which he displayed in figures such as Fig. 10.8 (see TP 36) in which AA' is parallel to B'C and AC is perpendicular to BC. Let BC $= a$, AC $= b$ as usual; BÂC $= P(\alpha)$, AB̂C $= P(y)$, Ĉ $= \pi/2$. Take AD $= x$, BD $= y$, and DD' perpendicular to AB and therefore parallel to AA'. Then A'ÂD $= P(x+y)$, but CÂD $= P(\alpha)$, so $P(x+y) + P(\alpha) = P(b)$. Similarly $P(x-y) = P(\alpha) + P(b)$.

A chain of similar arguments enabled Lobachevskii to establish that

$$P(b) = \tfrac{1}{2}\{P(x-y) + P(x+y)\}, \quad \cos P(b) = \cos \tfrac{1}{2}\{P(x-y) + P(x+y)\}$$

and

$$P(\alpha) = \tfrac{1}{2}\{P(x-y) - P(x+y)\}, \quad \cos P(\alpha) = \cos \tfrac{1}{2}\{P(x-y) - P(x+y)\}.$$

Therefore, since

$$\frac{\cos\{P(b)\}}{\cos\{P(\alpha)\}} = \cos P(x)$$

because of the interpretation of the spherical trigonometrical formulae

$$\frac{\cos\{P(b)\}}{\cos\{P(\alpha)\}} = \cos P(x) = \frac{\cos \tfrac{1}{2}\{P(x-y) + P(x+y)\}}{\cos \tfrac{1}{2}\{P(x-y) - P(x+y)\}}$$

and so

$$\tan\left\{\frac{P(x)}{2}\right\}^2 = \tan\left\{\frac{P(x-y)}{2}\right\} \tan\left\{\frac{P(x+y)}{2}\right\}.$$

However, the only function satisfying $\{f(x)\}^2 = f(x-y)f(x+y)$ is $f(x) = e^{-x}$ for some constant $e > 1$ so it must be that

$$\underline{\tan\left\{\frac{P(x)}{2}\right\} = e^{-x}}$$

which is the fundamental formula of non-Euclidean geometry (TP36, end).

A number of reformulations of this basic identity were given by Lobachevskii (TP37 end) and are of some interest. Since

$$\tan \theta = \frac{2 \tan (\theta/2)}{1-\tan^2 (\theta/2)} \quad \text{for all } \theta$$

$$\tan P(x) = \frac{2e^{-x}}{1-e^{-2x}} = \frac{2}{e^x - e^{-x}} = \frac{1}{\sinh x}.$$

$\cos P(x)$ can be found by rewriting

$$\tan \theta = \frac{\sin \theta}{\cos \theta} = \frac{1}{i} \frac{e^{i\theta} - e^{-i\theta}}{e^{i\theta} + e^{-i\theta}} = e^{-x}$$

where $2\theta = P(x)$ as

$$e^{-2i\theta} = \frac{1-ie^{-x}}{1+ie^{-x}}$$

so

$$e^{2i\theta} = \frac{1+ie^{-x}}{1-ie^{-x}}$$

so

$$\cos 2\theta = \frac{e^{2i\theta} + e^{-2i\theta}}{2} = \frac{e^x - e^{-x}}{e^x + e^{-x}} = \frac{\sinh x}{\cosh x}$$

or

$$\left.\begin{array}{l}\cos P(x) = \tanh x \\ \sin P(x) = \dfrac{1}{\cosh x}\end{array}\right\}. \tag{10.1}$$

They permit an interpretation of the fundamental formulae of non-Euclidean trigonometry to be made which brings Lobachevskii's formulae into the same form as those derived earlier by Taurinus. For, having established the formulae for a right-angled triangle, it is not difficult to establish the formulae which describe a general triangle, as Lobachevskii did (TP 37). Typical of his results are (TP 37, eqns 8):

$$\sin A \tan P(a) = \sin B \tan P(b)$$

and

$$\cos A \cos P(b) \cos P(c) + \frac{\sin P(b) \sin P(c)}{\sin P(a)} = 1$$

the latter of which Bonola selects as fundamental. Rewriting them as suggested they become, respectively,

$$\frac{\sin A}{\sinh a} = \frac{\sin B}{\sinh b}$$

and

$$\cos A \tanh b \tanh c + \frac{\cosh a}{\cosh b \cosh c} = 1$$

which reduces to

$$\cos A \sinh b \sinh c + \cosh a = \cosh b \cosh c$$

which, as we have seen, was used earlier by Taurinus. Incidentally, neither Lobachevskii nor Bolyai mention Taurinus at all, and they seem not to have known of his work.

Lobachevskii concluded his paper, except for establishing the formula (10.1), with the remark that:

The equations (8) attain for themselves already a sufficient foundation for considering the assumption of imaginary geometry as possible. Hence there is no means, other than astonomical observations, to use for judging the exactitude which pertains to the calculations of the ordinary geometry.

This exactitude is very far-reaching, as I have shown in one of my investigations, so that, for example, in triangles whose sides are attainable for our measurement, the sum of the three angles is not indeed different from two right angles by the hundredth part of a second.[4]

Janos Bolyai

By 1823 Bolyai had obtained the formula

$$e^{-a/K} = \tan \frac{P(a)}{2}$$

for the angle of parallelism, which expression as we have seen is the key to the non-Euclidean geometry.

The work continued until it was published at the son's expense in 1831, although father and son were not altogether happy with it even then because of the appearance in it of the unintelligible constant K in the formula above. The accident of Gauss's reception dissuaded the son from ever publishing again, although he continued to work on geometry until his death in 1860. The difference between the researches of Bolyai and of Lobachevskii is slight, the main difference being Bolyai's preference for working more directly with results which are absolute.

After some work he arrived at the following absolute definition of the horocycle and the horosphere, called by him L and F respectively (*Science absolute of space* 1831, § 11, to be referred to as SA): given a point A on a

---

[4] Lobachevskii's empirical attitude to the nature of space is discussed by Daniels (1975).

line AM, B lies on L (or in three dimensions upon F) if and only if it is the point on BN, a parallel to AM, such that NB̂A = MÂB.

The work of Bolyai's part is devoted to showing that the locus L of all possible B's exists absolutely. It is, of course, the locus of points corresponding to A with respect to a pencil of parallel lines (Gauss), and this property of chords to a horocycle was also known to Lobachevskii.

Familiar theorems followed:

(1) No three points of L, or F, are collinear (SA §16).
(2) Any plane meeting F does so in an L if that plane contains an axis (e.g. AM) and otherwise in a circle (SA § 18).
(3) The perpendicular BT to an axis BN of L is tangent to L (SA § 19).

He then easily obtained the following.

(4) Any two points of F determine an L and, by (4) above, the angle formed by L_1 and L_2, intersecting L's in F, is also the angle formed by the two planes which meet F in L_1 and L_2. Accordingly the angle sum of a triangle on F is π (SA § 20).

This means that as a theorem of absolute geometry, the geometry which obtains on F is Euclidean (SA § 21).

At this point, and in a manner so similar to Lobachevskii's as not to need repetition, Bolyai derives the formula $s' = se^{-x}$ (see above). The arbitrary constant K arises in our choice of e.

He next derived the absolute nature of formulae in spherical trigonometry (SA § 26) and then, by using the methods of calculus, the lengths and areas of familiar figures. These results of his are not found in Lobachevskii's *Theory of parallels*, although they are elsewhere, or in such detail in Taurinus, and a discussion of them is deferred till later. They mark an important change in attitude to the study of geometry, akin to the work of Gauss on general surfaces.

All of his formulae, however, involve an arbitrary constant K which arises in the manner indicated above. Bolyai comments (SA § 33) that, in the event of non-Euclidean geometry being real, a single measurement will determine the constant. Otherwise, there is a continuum of non-Euclidean geometries, alike unreal but intelligible to the mind, differing between themselves only with regard to the size of the constant. In an exactly similar way there is only one spherical geometry possible, but the measurements in it depend upon the radius of the sphere. Bolyai, however, could find no intuitive meaning for his arbitrary constant.

(In spherical geometry the area of a triangle is proportional to its angular excess $Â + B̂ + Ĉ - \pi$. The constant of proportionality is the square of the radius of the sphere. Thus the two triangles ABC have the same excess but, obviously, different areas.)

Finally he concluded by solving in non-Euclidean geometry one of the classical problems of geometry: squaring the circle, i.e. constructing a square equal in area to a given circle (SA § 43). At the time it was not known whether this could be done in Euclidean geometry, the first conclusive proof of impossibility being given in 1882, when Lindemann proved π to be transcendental. Oddly enough, an argument using infinite series to show that π is irrational was first given by Lambert,[5] who based his argument on Euclid X, 2. However, the irrationality of π is not the same as its geometric non-constructibility. Bolyai indeed pointed out that his proof will not work in the case of Euclidean geometry. If we indicate how his proof works you will see why.

The area of a square depends upon the angle in it, the area of a polygon being also proportional to its angular defect as we saw earlier (p. 68). We can[6] take the area of a circle as πC, so the desired square satisfies $\pi C = (2\pi - 4\alpha)C$, i.e. $\alpha = \pi/4$. In this fashion we have reduced the problem to that of constructing a triangle ABC which is one-eighth of the desired square. Whether such a construction can be performed depends upon whether a segment of length a which forms the base of the triangle can be constructed.

Now by the trigonometrical formulae obtained earlier

$$\cosh \frac{a}{K} = \frac{\cos(\pi/8)}{\sin(\pi/4)} = \frac{\sin(3\pi/8)}{\sin(\pi/4)}$$

which is perhaps more familiar on writing

$$\frac{\pi}{4} = 45°, \quad \frac{\pi}{8} = 22°30', \quad \frac{3\pi}{8} = 67°30'.$$

However, given an acute angle bounded by any two lines a and b, we may construct a line l which is parallel to a and perpendicular to the other, as follows.

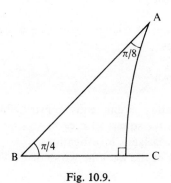

Fig. 10.9.

---

[5] J. H. Lambert, Mémoire sur quelques propriétés remarquables..., *Opera Math.* II, 2, (1761, 1768, 1767). An extract can be found in Struik (1969).
[6] A special case of the general formula area $= 2\pi K^2 \{\cosh r/K - 1\}$ (cf p. 95).

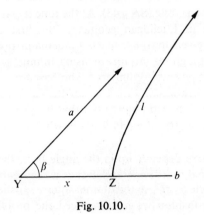

Fig. 10.10.

The segment $YZ = x$ is unique, and will be called the segment corresponding to β, where β is the angle of parallelism corresponding to x and

$$\cosh \frac{x}{K} = \frac{1}{\sin P(x)}$$

Construct therefore the segments corresponding to $3\pi/8$ and $\pi/4$, say b' and c', and note that

$$\cosh \frac{b'}{K} = \frac{1}{\sin (3\pi/8)}, \quad \cosh \frac{c'}{K} = \frac{1}{\sin \pi/4}.$$

Therefore

$$\cosh \frac{a}{K} \cosh \frac{b'}{K} = \cosh \frac{c'}{K}.$$

However, in the right-angled triangle with side b' and hypotenuse c' the remaining side a' satisfies

$$\cosh \frac{a'}{K} \cosh \frac{b'}{K} = \cosh \frac{c'}{K}$$

and so $a' = a$. This triangle is eminently constructible, and so using it to construct a we construct the square of area π.

This largely concludes Bolyai's investigations.

In the final paragraph of the work he wrote of Euclidean geometry Σ and the alternative geometry S:

It remains, finally . . . to demonstrate impossibility (apart from any supposition) of deciding *a priori* whether Σ or some S (and which one) exists. This, however, is reserved for a more suitable occasion.

This work was not attempted by him, so far as is known.

Summary

What Bolyai and Lobachevskii claimed is that mathematically a non-Euclidean geometry is possible. Both argued to this conclusion in the same way.

Parallels are defined as distinguished lines of a pencil, and elementary results are proved about them. With respect to a pencil of parallel lines a distinguished locus or surface is defined—the horocycle and the horosphere. Upon the horosphere geometry is Euclidean, this result being absolute. In space or in the plane formulae are given for a geometry which is non-Euclidean, the results of which are obtained via the formula of (absolute) spherical trigonometry. The bold step from studying plane geometry to studying solid geometry, so often left undiscussed in text books, is crucial for it ends the ambiguous search for a non-Euclidean 'plane' in a Euclidean three-dimensional space in which, almost inevitably, Euclidean concepts would continue to dominate the mind. The work of Bolyai and Lobachevskii thus takes implicitly an intrinsic approach to geometric concepts.

However, this does not conclusively establish the mathematical existence of non-Euclidean geometry; if it were to turn out that the only parallels were Euclidean ones the horocycle would collapse to the straight line, the horosphere to the plane, and the formulae to analytic truths descriptive of nothing. All that is established is that if you assume a non-Euclidean geometry to exist, it does so in a comprehensible way, as indicated by the precise formula for the asymptotic behaviour of two parallel lines. Indeed, Stäckel (quoted by Bonola, p. 112) tells us that Bolyai was still willing to contemplate finding a contradiction in non-Euclidean geometry.[7]

The equations of non-Euclidean trigonometry none the less seem to suggest sufficient grounds for believing an imaginary geometry to be possible. The difficulty is that from a false initial assumption true results may yet be obtained. Accordingly, the analytic formulae cannot be considered conclusive. Evidently, however, a self-contradiction in non-Euclidean geometry would imperil the whole of analysis as well—a prospect not easily to be faced. It was the achievement of the next generation of mathematicians to solve this dilemma in a twofold effort which was one part logic and one part purely mathematical.

Priorities

The question of priorities should be commented upon. It seems that Gauss was the first to conceive a non-Euclidean geometry, but he nowhere studied it as systematically as did Lobachevskii or Bolyai. Perhaps to a mind of his brilliance the formulae came so naturally that he felt able to restrict himself to commenting on and extending the work of his friends. There seems little reason to suppose that Lobachevskii knew any details of what was in Gauss's

---

[7] Bolyai was misled by a mistake in his calculations concerning the distances between five coplanar points.

mind until his work was largely finished. Any contact would have been through Gauss's friend Bartels, who stayed with him in 1807 and maintained a correspondence with him until 1817. As late as 1823, however, Lobachevskii was willing to consider attempts to prove the parallel postulate possible for he submitted a manuscript to St. Petersburg on elementary geometry in which he remarks to that effect. So the example of Gauss, if indeed he knew of it, cannot have counted for much.[8] He did, however, successfully propose Lobachevskii's election to the Göttingen Scientific Society in 1842.

As for Janos Bolyai, whatever the extent of communication between his father and Gauss, he was so outraged by Gauss's claim to know already the contents of his work that it does not make sense to imagine that he had prior acquaintance with any of Gauss's ideas. We may conclude that the discoveries were independent of each other, and we must look elsewhere for a historical explanation of how they came about. We shall see, in fact, that the real historical puzzles are much more interesting and complicated than mere priorities.

Farkas Bolyai was elected a corresponding member of the mathematics section of the Magyar Academy in 1832. He and his son unsuccessfully entered an international competition to explain the nature of complex numbers in 1837—no-one won. He died on 20 November 1856. It was his wish that his grave bear no mark. His son died four years later on 27 January 1860. A monumental stone was erected upon the grave in Maros-Vasarhely by the Hungarian Mathematico-Physical Society in 1894.

Lobachevskii, with his unpopular liberal opinions, was never appreciated in his lifetime and died in poverty. Scarcely a year before he died he published, in Russian and French, his third full-blooded attempt to get the new geometry known. He was by then blind. He died on 24 February 1856; the centenary was marked by a publication by the University of Kazan of a book devoted to his work.

Carl Friedrich Gauss died in 1855, the Prince of Geometers. His notebooks revealed more mathematics that, with his massive conservatism, he had not dared to publish.

Exercises

10.1 Check the steps in Bolyai's construction of a square of area π (p. 109). Can you see how to perform the constructions which are there taken for granted?

10.2 Prove the 'prism theorem' (p. 99).

---

[8] May (1972) considers his influence to have been, on the whole, retrogressive. Biermann (1969) finds no evidence of any influence in the Gauss–Bartels correspondence. *Naturwissenschaft, Tech. Med.* **10**, 5–22.

Appendix

Spherical trigonometry

The general formulae for *plane* trigonometry can be derived from the special formulae which apply only to a right-angled triangle because any triangle can be divided into two right-angled triangles. The same approach can be made to yield the formulae of spherical trigonometry.

Consider first the case of a spherical triangle with a right angle at A. Cartesian co-ordinates can be chosen so that A is at $(1, 0, 0)$, B is at $(\cos c, \sin c, 0)$ and C is at $(\cos b, 0, \sin b)$. The radius of the sphere is taken to be 1, so $AB = c$ and $AC = b$. We first of all find BC, which is, of course, equal to the angle COB.

If the chord length $BC = x$, then $x^2 = (\cos b - \cos c)^2 + \sin^2 c + \sin^2 b$ by Pythagoras theorem, so $x^2 = 2 - 2\cos b \cos c$, but in the great circle centre O through B and C we have $x/2 = \sin(a/2)$, so

$$x^2 = 4\sin^2 \frac{a}{2} = 2(1 - \cos a)$$

$$\underline{\cos a = \cos b \cos c.} \qquad (10.2)$$

We next find the angle $\hat{A}$. The plane through O, A, and B has equation $z = 0$. The plane through O, B and C has equation $x - y \cot b - z \cot c = 0$. The angle $\hat{B}$ between these planes, is given by

$$\cos^2 \hat{B} = \frac{\cot^2 c}{1 + \cot^2 b + \cot^2 c}.$$

The formula we seek

$$\underline{\cos b = \cos c \cos a + \sin c \sin a \cos \hat{B}}$$

can now be derived by an unedifying slog using (10.2). Similarly we can deduce

$$\underline{\cos c = \cos a \cos b + \sin a \sin b \cos \hat{C}.}$$

In the same way we next deduce the formula

$$\underline{\cos \hat{A} = -\cos \hat{B} \cos \hat{C} + \sin \hat{B} \sin \hat{C} \cos a}$$

which, since $\hat{A} = \pi/2$ reduces to

$$\underline{\cos \hat{B} \cos \hat{C} = \sin \hat{B} \sin \hat{C} \cos a.}$$

Observe that it is enough to establish that

$$\cos^2 \hat{B} \cos^2 \hat{C} = \sin^2 \hat{B} \sin^2 \hat{C} \cos^2 a$$

i.e. that

$$\cos^2 \hat{B} \cos^2 \hat{C}(1 - \cos^2 a) = (1 - \cos^2 \hat{B} - \cos^2 \hat{C}) \cos^2 a$$

Exercise

Check that for example

$$\cos \hat{B} = -\cos \hat{C} \cos \hat{A} + \sin \hat{C} \sin \hat{A} \cos b$$

which of course reduces to

$$\cos \hat{B} = \sin \hat{C} \cos b.$$

Finally we check that

$$\frac{\sin a}{\sin \hat{A}} = \frac{\sin b}{\sin \hat{B}} = \frac{\sin c}{\sin \hat{C}}$$

which in our case reduces to checking $\sin a = \sin b/\sin \hat{B}$ and is again true.

We now deduce the results for a general spherical triangle from the results for a right-angled one by regarding the general triangle as being made up of two right-angled ones. Take a triangle $\tilde{B}CD$ which is split into two right-angled triangles $A\tilde{B}C$ and $A\tilde{B}D$ by dropping the perpendicular $\tilde{B}A$ to DC. We have

$$\widehat{DBA} = \hat{B}', \quad \widehat{DA} = b', \quad \widehat{ABC} = \hat{B}$$
$$AC = b, \quad \widehat{BD} = a', \quad \widehat{DBA} = \hat{B} + \hat{B}'.$$

We must establish, for example, that

$$\cos(b+b') = \cos a' \cos a + \sin a' \sin a \cos(\hat{B}+\hat{B}')$$

from the above formulae and others like them, such as $\cos a' = \cos b' \cos c$, derived from the triangle $D\tilde{B}A$.

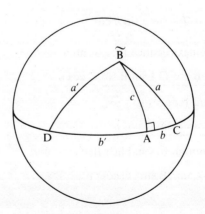

Fig. 10.11.

LHS $= \cos b \cos b' - \sin b \sin b'$

RHS $= \cos b \cos^2 c \cos b' + \sin a \sin a'(\cos \hat{B} \cos \hat{B}' - \sin \hat{B} \sin \hat{B}')$

$= \cos b \cos b' \cos^2 c + \sin a \sin a' \cos \hat{B} \cos \hat{B}' - \sin b \sin b'$

so it is enough to show that

$\cos b \cos b' = \cos b \cos b' \cos^2 c + \sin a \sin a' \cos \hat{B} \cos \hat{B}'$

i.e. that

$\cos b \cos b' \sin^2 c = \sin a \sin a' \cos \hat{B} \cos \hat{B}'$

i.e. that

$\cos b \cos b' \sin^4 c = (\sin a \sin c \cos \hat{B})(\sin a' \sin c \cos \hat{B}')$

which is true since

$\cos a = \cos b \cos c, \quad \cos a' = \cos b' \cos c.$

The formulae of spherical trigonometry hold absolutely.

To give a flavour of J. Bolyai's work, this is a summary of his proof which is easier than Lobachevskii's. He has already laid down the necessary definitions and theorems concerning horocycles L and horospheres F, and shown that the geometry on the horosphere is Euclidean. Upon the HAA he next showed

(1) in any rectilineal triangle ABC if we write $C(a)$ for the circumference of the circle of radius a, etc., then

$C(a):C(b):C(c) = \sin \hat{A} : \sin \hat{B} : \sin \hat{C};$

(2) in any spherical triangle ABC

$\sin a : \sin b : \sin c = \sin \hat{A} : \sin \hat{B} : \sin \hat{C}$

which is the desired result.

His argument to show (1) is essentially a prism argument (SA25). It is enough to establish it for a right-angled triangle ABC, with the right angle at C, imagined lying on the ground as in Fig. 10.5. Take the straight line AA' perpendicular to ABC, and the parallels BB', CC' to AA'. Since $\hat{C} = 90°$ the walls of the prism AA'CC' and CC'BB' are also at right angles to each other. Now consider the horosphere defined by the three parallels and passing through A (Bolyai makes it pass through B, but it makes no difference). It meets BB' at D and CC' at E say. As you expect, $D\hat{E}A = 90°$.

This is sufficient. Since BDE is a Euclidean triangle the formulae of plane trigonometry apply and we deduce

$AD:1 = DE:\sin D\hat{A}E, \quad \text{so } AD:DE = 1:\sin D\hat{A}E.$

However, $\widehat{DAE} = \widehat{BAC}$, and

$$\frac{AD}{AE} = \frac{C(AD)}{C(AE)}$$

since the geometry on the horosphere is Euclidean, and

$$\frac{C(AD)}{C(AE)} = \frac{C(BA)}{C(CA)}$$

since the plane ABC meets any horosphere other than those with AA' etc. as axes in arcs of circles.

To prove (2), Bolyai argued as follows (SA § 26):

"For take $A\hat{B}C$ = a right angle, and CED perpendicular to the radius OA of the sphere. We shall have CED perpendicular to AOB, and (since BOC is perpendicular to BOA), CD perpendicular to OB. But in the triangles CEO, CDO (by § 25) C(EC):C(OC):C(DC) = sin $C\hat{O}E$:1:sin $C\hat{O}D$ = sin AC:1:sin BC; meanwhile also (§ 25) C(EC):C(DC) = sin $C\hat{D}E$:sin $C\hat{E}D$; but $C\hat{D}E$ = right angle, and $C\hat{E}D$ = $C\hat{A}B$. Consequently sin AC:sin BC = 1:sin A."

This is the desired conclusion.

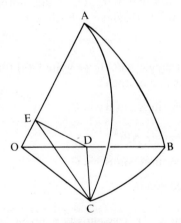

Fig. 10.12.

11 Curves and surfaces

Thirty years were to pass before the impact of the work of Lobachevskii and Bolyai was to be adequately felt; the new work was not to attract mathematicians until it was made easier to understand. The analytical techniques of the pioneers lacked the power to convince, and in the end it was the work of Riemann and Beltrami which finally established non-Euclidean geometry as a valid object of study, and in so doing changed our understanding of geometry and the world. The climate of ideas that they created eventually made relativity possible.

No one would dispute that Bernhard Riemann is incomparably the greater of the two, for he is one of the true fathers of modern mathematics. In 1854 he presented his celebrated memoir, *On the hypotheses which lie at the foundations of geometry*, to a general audience at Göttingen. Gauss had selected the topic as part of Riemann's examination for his Habilitation, and he later told W. Weber how impressed he was with the young man's profundity (Freudenthal 1975). To a modern reader, however, the difficulties of the paper are heightened by its presentation in a non-technical fashion. We can best approach it by discussing the work which led up to it.

Curves

There are innumerable curves one can draw on paper, of which only a few have names, e.g. the circle, the ellipse, the cycloid, and various epicycloids. Some of these are related to others by means of simple constructions; thus a cycloid is described by a point on the rim of a wheel rolling along a straight line. In order better to study curves in general the first step was to parametrize them, to associate to each point on the curve a name, and very often an explicit parallel is made with the description of the curve in time (the name given to a point is the time at which the drawing instrument reached it). In this fashion we may parametrize arbitrary curves, and since t, the parameter, may take positive or negative values we often consider the curve as a map from the real numbers to the plane.

The same curve may be parametrized in many different ways, according, for instance, to the speed of the drawing instrument, and so one must distinguish properties of the curve (which are independent of the choice of parameter) from accidents arising from a particular choice of parameter. Here we shall not allow these details to detain us, but merely note the problem. It will be discussed later.

Space curves

Curves may also be drawn in three-dimensional space, and following the lead of Clauraut's *Recherches sur les courbes à double courbure* (1731) Euler and others made a thorough study of space curves during the eighteenth century. A point on a curve in space can be parametrized by some parameter, say t, and located with reference to the usual co-ordinates x, y, and z (denoting 'along', 'away', and 'up' from some arbitrary fixed origin).

We write $P(t)$ for the position of the tracing point at time t, and give it co-ordinates $x(t)$, $y(t)$, $z(t)$ if we want to be precise. Obviously as t changes so do x, y, and z and consequently we have an exact description of the curve. We have a ready picture of the tangent to the curve at P as indicating in what direction the point is moving at time t. Although not tracing out a circle, to a good approximation near P the point travels on a sphere. A unit line along the radius drawn inwards from P will be called principal normal to the curve at P. In the case of plane curves the situation is as follows.

The circle S closely approximates the curve at P. It is called the circle of curvature and OP is called the radius of curvature of the curve at P; curvature is defined, perversely,[1] as the reciprocal of the radius of curvature and is usually denoted by κ. Knowing the curvature of the curve tells you in what way it departs from the tangent at that point. The three-dimensional picture is analogous; however, some measure of how the curve departs from the plane of the normal and the tangent is required (see Fig. 11.2). Such a measure is called the torsion, denoted τ, and is measured along a line perpendicular to that plane.

Monge[2] and Lancret set out the first good approach to space curves, and the exposition in terms of the description we now use was first given by Frenet (in 1847) and Serret (in 1851). It was found that by considering only the curvature and torsion at points on the curves a description of the whole

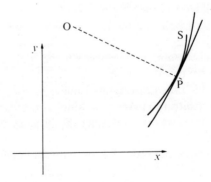

Fig. 11.1.

---

[1] The more 'curved' a sphere is the smaller it is, i.e. the smaller its radius is.
[2] See C. Monge, *Applications de l'analyse à la géométrie* (1807).

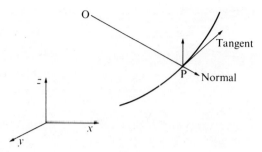

Fig. 11.2.

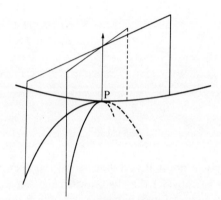

Fig. 11.3.

curve can be obtained; for example, curves for which κ/τ is constant are helices.[3]

Surfaces

From curves it is a natural step to consider surfaces, and the transition was made in two ways. One way was via the study of geodesics (see p. 124). The question is: What is the shortest path on the surface between two given points? The answer characterizes the shortest path as special in some way amongst paths on the surface; a string stretched tight along the surface takes up the geodesic position, essentially because there are no sideways forces upon it. An alternative approach to surfaces is to take cross-sections by means of planes and so obtain curves. This was the approach of Euler in work done from 1760 onwards,[4] Euler took just those planes which contained the normal

---

[3] A theorem stated by Lancret, a pupil of Monge, in 1802, and first proved by B. de Saint Venant, *J. Ec. Polytech.* **30**, 26 (1845).

[4] For example L. Euler, Recherches sur la courbure des surfaces *Acad. Sci. Berlin*, pp. 119–143 (read 1760, published 1767).

120 *Curves and surfaces*

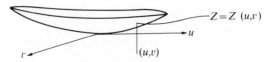

Fig. 11.4.

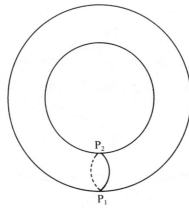

Fig. 11.5.

to the surface at a given point P and considered the curves so obtained. It was left to Meusnier to consider the algebraically messier problem of cutting the surface with any plane. Euler particularly studied the curvature of the 'slice' curves that are made on the surface in this fashion. He found that of these curves through a point P on a given surface there is, in general, precisely one of the greatest curvature and one of least curvature. These curves are called the principal curves through P. Furthermore they meet at P in a right angle. This result may well surprise, and I shall discuss it in manner of Meusnier[5] whose proof is more elegant and modern than Euler's.

Take a point P on the surface and look at the plane tangent to the surface at P. We may give this plane (u, v) co-ordinates, and we then express the equation of the surface near P in terms of its height above or below this tangent plane. Meusnier showed that if r is the radius of curvature of the surface at P, informally the radius of the sphere which most closely approximates the surface at P, then the new surface generated by rotating a part of the circle drawn in the $v = 0$ plane, which has radius r and passes through P, has a remarkable property. It has the same principle curvatures as the original one, provided a certain harmless condition is satisfied. The new surface is a portion of a circle swung along a line and so is part of a torus (see Fig. 11.5). There are on a torus essentially two types of points: the base

---

[5] G. Meusnier, 'Mémoire sur les courbure des surfaces', *Mem. Savants Étrangers*, **10**, 477–501 (read 1776, published 1785).

P_1 and the saddle P_2, to which our new surface conforms. At either of these, as at all points on the torus, the principal curvatures meet at right angles. Therefore, by Meusnier's argument, the principal curves on the original surface meet at right angles as well. Furthermore, we see that locally all (non-degenerate) surfaces look either like bowls, upward or downward facing, or saddles, since those are the range of possibilities on the torus.

Euler was also the first to notice that essentially a surface is described by only two parameters, but the full significance of this result was apparent only to Gauss.[6]

Co-ordinates on surfaces

We habitually view land in two ways: as a surface on which we walk about and which can be described with two co-ordinates (e.g. latitude and longitude), and as a surface which rises and falls and invokes a third dimension of height.

A rectangular grid on a flat surface could be made up of squares of equal area. However, if such a grid was to be draped over a hill it would have to be stretched until the former squares were not only deformed, but no longer of equal area. This is why Elastoplast which stretches is a better fit than ordinary Elastoplast for cuts on knuckles and knees.

Now if we were to put co-ordinates on a surface we could think of them as being like a net. Each point would have a two-number address, a value for u and v in Fig. 11.6, just as if stretchable graph paper had been draped over the surface. On the second way of seeing a hill, as a surface in a three-dimensional world, we would give points three co-ordinates, but only certain combinations would arise since x and y together determine z, the height.

The (u, v) description is *intrinsic*; it is the only description available to beings constrained to live in the surface. E. A. Abbott's 'Flatlanders' were

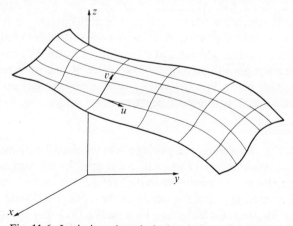

Fig. 11.6. Intrinsic and extrinsic descriptions of a surface.

[6] See L. Euler, *Opera posthuma*, Vol. 1, p. 494 (1862).

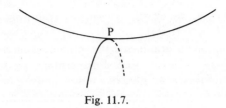

Fig. 11.7.

Euclidean, and one point of his charming tale[7] is to impress us with the intrinsic description we use of the three-dimensional world, but of course his aliens were three-dimensional, and their view of the two-dimensional surface is *extrinsic*, precisely because they stand outside it (above and below it). The surface is describable both intrinsically and extrinsically, and the tension set up between these two descriptions was to be richly exploited by Gauss.

The curvature of a surface at a point is the ratio $\lim_{A,B,C \to P}$ (area A'B'C' : area ABC) where ABC is a triangle enclosing P and A'B'C' is the triangle on the celestial sphere defined by the perpendiculars AA', BB', CC' to the surface. It is very easy to see extrinsically, and when we come to consider the curves formed by the method above an interesting connection turns up between the curvature and the principal curves. Apart from degenerate cases it turns out that the curvature K is equal to the product $k_1 k_2$ of the curvatures k_1 and k_2 of the principal curves. This is always true provided two principal curves exist (some typical degenerate cases are described in the Appendix to this chapter). For example, in the case of a cylinder, the diameters are circles and are of steepest curvature. The flattest curves are the longitudinal axes, which are straight lines. Therefore the curvature of a cylinder is zero and the cylinder is said to be bent but not curved, which is as it should be for rolling the cylinder across a flat inky sheet of graph paper would impart to it a rectangular grid which over any small part is indistinuishable from the flat grid. (In the large it overlaps itself, but that does not affect the local geometry.)

For a saddle, the curves indicated at P (see Fig. 11.7) are the principal curves and are curved in opposite directions. The centre of one is above P and the centre of the other below P, so k_1 and k_2 have opposite signs and the curvature of the saddle at P is therefore negative.

Curvature (intrinsic)

As the osculating sphere and the principal curves are all extrinsically determined, because their centres and radii do not lie in the surface, they are unintelligible to beings who only live in the surface. Yet we feel that curvature is somehow intrinsic, and that the nature of the (u, v) grid should reveal itself to our surface dwellers. Gauss[8] discovered before 1827 that this is indeed the

---

[7] E. A. Abbott, *Flatland—a romance of many dimensions—by a square* (1884).

[8] Gauss's fundamental work in this subject is his *Disquisitiones generales circa superficies curvas*, 1827; *Werke*, IV, 217–58.

case, in a theorem he stumbled on almost by accident having found it to be so in numerous examples (a common method of discovery for Gauss, who, like many great mathematicians, was addicted to the calculation of examples). So delighted was he with the theorem that he gave it a name, which it still holds, the *Theorema Egregium* or outstanding theorem. Mathematically, if curvature is to be intrinsic it must be capable of expression not just in terms of x, y, and z, but in terms of u and v alone. Notice that there is no hope that k_1 or k_2 can be written in that way. The *Theorema Egregium* shows that K can indeed be written in terms of u and v alone.

Beings intrinsic to the surface can determine the curvature in various ways. Imagine that they wish to lay down a square grid on a hill. If at each vertex they lay off a fixed distance at right angles to the last they get into trouble for they do not construct a square but an open figure. To close the square they must shrink one of the sides or change one of the angles, thus obtaining a figure reminiscent of the quadrilaterals of Saccheri and Lambert. The extent to which their 'square' does not close up is a measure of the curvature of the surface in the limit as steadily small sides are drawn.

An even deeper connection between the previous work on non-Euclidean geometry and the study of surfaces is obtained in the next theorem. It was obtained by Gauss and relates the curvature of a triangle to the angular defect of triangles drawn upon it. For a general surface the curvature K varies from point to point and so defines a function on the surface. By integrating this function over the region defined by the triangle ABC whose sides are geodesics, Gauss showed

$$\iint_\Delta K \, dS = \alpha + \beta + \gamma - \pi$$

where α, β, and γ are the angles at A, B, and C. In particular, when K is constant (the surface has constant curvature)

$$\iint_\Delta K \, dS = K(\text{area} \triangle \text{ABC}) = \alpha + \beta + \gamma - \pi$$

or

$$\text{area} (\triangle \text{ABC}) = \frac{\alpha + \beta + \gamma - \pi}{K}$$

which is the connection between area and angular defect, or excess, of a triangle on such a surface.

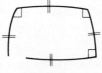

Fig. 11.8.

124 *Curves and surfaces*

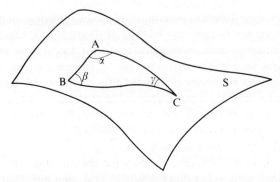

Fig. 11.9. The triangle ABC is composed of geodesics on the surface S.

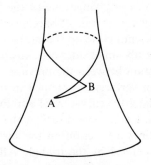

Fig. 11.10.

Gauss's work established the intrinsic language with which surfaces could be described and analysed. Other mathematicians were able to extend his techniques to novel problems, and in this way differential geometry began to grow. Of course, the simplest surface of all is the plane, and a concept so central to our understanding of plane geometry is the straight line. Naturally, we would hope to define an analogous concept for surfaces, and we can almost do this. We abstract from the planar situation the property of a straight line that it is the curve of shortest distance between the points it joins. On the sphere, for instance, the curve of shortest distance between two points is the section of the great circle that lies between them. On a more general surface, however, we cannot hope to be so lucky; we can look in vain for a curve of minimal length.[9] A plane, with a disc of radius one unit removed from its centre possesses no geodesics joining opposite points. Successively shorter paths can be found edging closer and closer to the edge of the pond, but there is no path of minimal length. Even if we exclude by arbitrary fiat such unpleasant surfaces as these, there is a second problem. To select the curve of

---

[9] This curve is called a geodesic. Apparently Liouville was the first to use the word in this sense (see Struik 1961).

minimal length from all the possible curves joining two points is a problem calling for the techniques of the calculus, which, as is well known, select maxima and minima only locally from among the nearby values. Thus the curve has a local minimum at $x = 1$ and a local maximum at $x = -1$, but clearly no overall minimum or maximum is ever obtained. In a similar way the search for geodesics on a surface, such as the one in Fig. 11.10, results in the curve AB being drawn as shown, but there are other geodesics, of which one is shown which wraps itself once around the surface and is minimal in length amongst such paths. We can only conclude that differential geometry is stranger than plane geometry.

Elastic bands provide a happy physical analogy with geodesics, an analogy which runs deep into mechanics which was a lively area of research in the nineteenth century. A band stretched between two points will always take up a configuration of minimum energy. On the plane it lies in a line, on the sphere it lies in a great circle. On more general surfaces it will try to lie along a geodesic, but notice that in the pond example it is impossible to stop it falling in. Taking an elastic band to the surface above will help to convince you that the wrap-around curve is a geodesic.

Minding's surface

An interesting surface with this property of multiple geodesics between points was discussed by H. F. Minding[10] in the 1830s and then largely forgotten for thirty years; this was a surface of constant negative curvature. Just as one cannot draw some curves in the plane without points of self-intersection arising (for instance knots) so similar surfaces can arise with surfaces drawn in three-dimensional space, and Minding's surface is conventionally drawn with a circle of singular points, points at which the picture is severely inaccurate. The surface is formed by rotating a tractrix about its vertical axis. A tractrix, the name is due to Huyghens,[11] is the curve of the obstinate dog. A point Q on a line l is attached to a point P by a curve of fixed length. P, the dog, is dragged behind Q, the owner, who walks along l, and the path of P is called

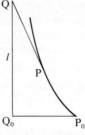

Fig. 11.11.

[10] H. F. Minding's articles appeared in the *Journal für Mathematik* for 1839 and 1840.
[11] In correspondence, February 1693; Letters no. 2793 and 2794, *Oeuvres*, Vol. 10.

126 *Curves and surfaces*

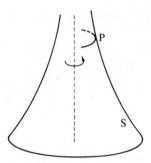

Fig. 11.12. Minding's surface.

the tractrix. It is conventional for the walk to start with PQ perpendicular to l, i.e. in position P_0Q_0 in Fig. 11.11. Minding's surface, which I denote S and which is sometimes called the pseudosphere, is formed by rotating the tractrix about l; P_0 generates a circle of singular points.

At any point, P, the surface looks locally like a saddle; the obvious line of steepest ascent is a principal curve and so is the horizontal circle through the point. They are therefore negatively and positively curved respectively, and their product, the curvature, is everywhere negative. It is furthermore everywhere constant since, informally anyway, as one rises up S the principal vertical curvature decreases and the principal horizontal curvature increases so as to preserve the value of the product, and moving sideways obviously changes nothing.

The geometry of Minding's surface is described by figures made up of geodesic segments—strings laid upon it and pulled taut. They must, of course, be constrained to lie always on the surface and not to jump across it like tunnels in the Earth. Minding's interest was in surfaces which, over small regions at least, were metrically the same like the cylinder and the plane, as we saw above, because one can be rolled across the other. He noticed that bending does not change the geometry but that stretching does, which is why paper crinkles if you try and wrap it round something curved and curtains hang only in folds. For a surface initially flat to fit onto a curved surface it may need to stretch and distort like skin does. He also noticed one property of the surfaces of constant curvature which will be important in what follows. Any figure drawn upon such a surface (plane, sphere, or pseudosphere) may be slid about without changing its shape. This too reflects the intimate connection between curvature and stretching. To see this, return to the argument earlier that showed that no square can be drawn on a curved surface—if the angles are right the edges do not join up. Now deform the figure you have made until the edges do join up, and try and move it around. If you think of curvature as a force which snapped open the original square, a change in curvature will now snap open the figure you are moving around. Therefore the only surfaces which admit the sliding of congruent figures are those of constant curvature.

(The tractrix and its axis resemble the picture of asymptotic parallels due to Saccheri and Gauss; this resemblance is not accidental. The surface S was also known to Gauss; he referred to it in an unpublished note written between 1823 and 1827 as the surface of revolution which is the opposite of the sphere but he does not appear to have connected it with anything in non-Euclidean geometry (*Werke*, VIII, p. 264).)

Appendix

Meusnier's theorem establishes that locally surfaces are mostly like either bowls or saddles. When the conditions of his theorem break down the shape of a surface near a given point can take other forms. Consider first the non-degenerate case. A plane parallel to the tangent plane at a point on a bowl meets the bowl in a closed curve which, to a good approximation, can be taken as an ellipse. Such points on a general surface are consequently called elliptic. Points on a saddle-like surface are called hyperbolic, because in their case planes parallel to the tangent plane meet the surface in (approximate) hyperbolae. There is an intermediate case of parabolic points, which are degenerate. On a torus the elliptic points are separated from the hyperbolic points by two circles of parabolic points which lie at the top and bottom if the torus is horizontal. There is only one surface in which every point is parabolic, and that is the sphere. Typically, the argument about principal curves breaks down at parabolic points, and indeed on the sphere every great circle through a point is a principal curve.

A more interesting kind of degeneracy is afforded by the monkey saddle on which three principal curves meet at the degenerate point. The equation of the surface is $z = x^3 - 3xy^2$; it derives its name from the evident fact that such a saddle has two downward parts for the legs and one more for the tail (Fig. 11.13).

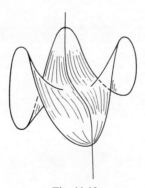

Fig. 11.13.

On the subject of geodesics as the correct definition of straight, it is interesting to note Heron's definition (*Metrica*, 4)

A straight line is a line stretched to the utmost

and also Galileo's comment in his *Dialogue concerning the two chief world systems* (trans. Stillman Drake, 1967, p. 16. University of California Press, Berkeley).

... simple lines, these being the straight and circular only, ... nor do I care to quibble about the cylindrical helix, of which all parts are similar and therefore seems to belong among the simple lines.

12 Riemann on the foundations of geometry

It is in this context that Riemann produced his memoir with the aim of clearing up the confused situation of geometry. I shall paraphrase his paper, adding illustrative comments in square brackets. Although what Riemann has to say is hard, its importance is considerable. For the first time it becomes possible to think geometrically in terms more basic than those of Euclid, with the result that ambiguities and difficulties in Euclid's formulation can be resolved. After Riemann it gradually became clear to mathematicians that Euclid had made many more assumptions than he had explicitly stated; in particular the assumption that any length can be doubled had greatly coloured his idea of geometry (see p. 150). Further, it became possible to design geometries that were highly non-Euclidean, lacking many properties of Euclid's but having new ones of their own, and these new geometries now turn up frequently in physics, notably in relativity theory. So let us look at this important paper. It is worth remarking that in this paper Riemann never mentioned non-Euclidean geometry by name but it is there by implication, an implication his colleagues were not slow to draw. He began with a reference to obscurities in the nature of space that geometers from Euclid to Legendre had failed to clear up, and for illustrative purposes he considered surfaces of constant curvature, where of course the surface of zero curvature is the plane possessed of Euclidean geometry, as he remarks. He also observed that if bodies can be moved around without distortion then

... it follows that the curvature is everywhere constant; and then the sum of the angles is determined in all triangles when it is known in one (*Hypotheses*, III, 1)[1]

which we have earlier called the three-musketeers theorem (p. 56).

Geometry, he remarked, always presupposes some fundamental notions for constructions in space which are undefined and which relate to one another in a way laid down by axioms. [Thus point and line are undefined in familiar geometry, and an axiom is that given any two points there is a unique line joining them.] The relationship between these presuppositions remains in the dark: 'one sees neither whether and in how far their connection is necessary, or *a priori* whether it is possible'. He proposed to discuss space, by which he meant the world in which we live, as one particular case of a 'multiply extended magnitude' constructed out of a general notion of quantity. It turns out, he said, that such magnitudes are capable of various metrical relations one of which will correspond to the world we live in. Which one it is 'can be gathered only by experience'. (*Hypotheses*, Plan.)

[1] References to the paper will be given thus: *Hypotheses*, III, 1, refers to paragraph 1 of chapter III of the paper. It was published in the Abhandlungen K. Ges. Wiss. Göttingen, vol. 13, 1867, and in Riemann *Werke* (1876, Dover 1953), 272–287.

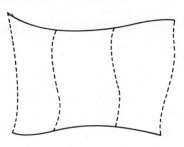

Fig. 12.1.

An n-fold extended magnitude, or manifold, would arise in this way. A concept 'whose mode of determination varies continuously' [colour or distance along a line] if one passes in a definite way from one mode to another constitutes a simply extended manifold (e.g. the spectrum). A transformation taking each point of such a one-manifold to a point of another would form a two-fold magnitude and so on. In the one-dimensional case only motion forforward and back is possible, two different directions are needed to describe motion on a 2-manifold, and so on. 3-manifolds and more generally n-manifolds may be introduced likewise.

We may also run the construction backwards, extracting $(n-1)$-manifolds from n-manifolds. [Here the appropriate picture is contour lines or level lines on a map. A surface of a hill is a 2-manifold, and the contour lines are curves and as such 1-manifolds.] (*Hypotheses*, I, 3.) Riemann was even willing to contemplate manifolds of infinite dimension, and this at a time when dimensions greater than three were seldom considered. We need not follow him here.

He next turned to the problem of introducing relations of measure on an n-manifold; we follow using throughout the idea of a 3-manifold, of which Riemann says space is one. We shall consider magnitude to be independent of location (*). A point has co-ordinates x_1, x_2, x_3 and a curve on a 3-manifold is (see p. 118) a parametrized set of points $(x_1(t), x_2(t), x_3(t))$ for some range of values t. Moving along a curve continuously changes t and dt and each x by dx [the language of infinitesimals in his].

Under reasonable conditions the square of the line element, the distance ds^2 along the curve, is a quadratic function of dx_1, dx_2, and dx_3. For instance, it might be that $ds^2 = dx_1^2 + dx_2^2 = dx_3^2$, a 'Pythagorean' situation, and in general the function will be everywhere positive. (Accordingly we can take square roots without fear of introducing 'imaginary' distances). We have in this way given a foundation to the way a threefold extended magnitude (extended along x_1, x_2, and x_3) allows the determination of the length of curves.

Suppose next we pick a point and draw the shortest paths out of it for a little way. Any nearby point can now be specified, as it were radially, by asserting how far along it lies upon one of these lines. (*Hypotheses*, II, 2.)

These 'polar' co-ordinates can be changed into our original x_1, x_2, x_3 co-ordinates,[2] and the square of the line element can be recaptured in a way which involves dx_1, dx_2, and dx_3 and perhaps x_1, x_2, and x_3. [See below for the two-dimensional case (x, y) or (r, θ).] If geodesics are not curved with respect to the co-ordinates, no terms involving x_1, x_2, or x_3 will appear in the expression for ds^2 (e.g. in Euclidean 3-space $ds^2 = dx_1^2 + dx_2^2 + dx_3^2$ shortest lines look 'straight').

[The assumption (*), that a segment can be moved around without changing its length, is much weaker than the assumption that a body can be thus transported for it admits geometries on surfaces which are not of constant curvature.]

Suppose one considered a different manifold, say a 2-manifold or surface, carrying a different metric. The terms involved in the expression for ds^2 at a point are related to Gauss's curvature (*Hypotheses*, II, 3). In particular, he remarks, for surfaces of constant curvature motion of fragments is possible without stretching—even without bending—in the case of the plane and the sphere but not in the pseudosphere (*Hypotheses*, II, 4).

Riemann concluded with a brief 'Application to space'. If curvature is everywhere zero we have Euclid's geometry. If, however, one assumes that figures can be moved without harm we must inhabit a world of constant curvature, which can be determined by consideration of the angle sum of a triangle.

He distinguished the unbounded from the infinite. Just as a curve can be without end (of unlimited extent) but finite in magnitude—a circle for instance—so space can be unlimited, but not necessarily infinite. Indeed, it might well have constant positive curvature and therefore finite radius (*Hypotheses*, III, 2).

The paper concludes with a summary of its method, a transition from the small to the large, from the local properties of extended magnitudes to the global properties of the manifolds so described, and explicitly from mathematics to physics.

What is Riemann saying? Let us discuss surfaces from his point of view; the higher-dimensional case will be taken up in the final part of this book. A surface is a 2-manifold, since at any point P in Fig. 12.2 two basic directions must be specified and then any third direction is some combination of those two. These directions can be taken as the axes of a co-ordinate system near P. There is no uniqueness in our choice of these directions, since not only would any two others do just as well but an entirely different co-ordinate grid could be put on the surface near P.

Two familiar grids on the plane are the rectangular grid and the polar grid. The polar one is made up of lines radiating from some fixed point (the origin)

---

[2] Note that x_1, x_2, x_3 are names for co-ordinate axes, and not the location of any particular point.

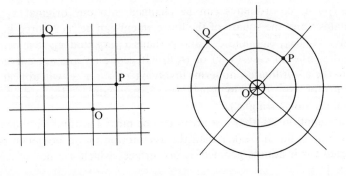

Fig. 12.2.

and circles centre the origin, and so, unlike the rectangular grid, the origin is a special point. We shall take our point P to be anywhere other than the origin.

Riemann next discussed ways of measuring distance on a surface, which implicitly involves some heavyweight calculus. For our purposes it is enough to know that the length along a curve is somehow determined, so that in particular the shortest curve or geodesic between P and Q can be found. We saw above that this cannot always be done; we give ourselves the ability to do this at least for all nearby P and Q. With the familiar metric on the plane (but there are others) geodesics are straight lines, and in this way we can distinguish between the two co-ordinate grids. In the rectangular grid every co-ordinate line is a geodesic between the points on it, but not every co-ordinate line in the polar case is: the radial ones are, but the concentric circle are not.

The business of the ds^2 and the dx_i^2 comes down to this. In the rectangular case an infinitesimal move from P to Q which is made up of a move of dx_1 along x_1 and dx_2 along x_2 results in a move of $ds = \sqrt{(dx_1^2 + dx_2^2)}$ along the hypoteneuse. The Pythagorean formula is specially simple because the grid is made up of geodesics. However, in polar co-ordinates a move of dr radially and then $d\theta$ circularly goes from $P = (r, \theta)$ to $Q = (r+dr, \theta+d\theta)$. The second move is through a distance of $(r+dr)(d\theta)$ and the distance PQ after a little computation is found to be approximately $ds = \sqrt{(dr^2 + r^2\, d\theta^2)}$.

The ambiguous phrase 'geodesics curved with respect to the co-ordinates' means only that if travelling circularly is 'natural' then travelling geodesically will feel unnatural. This is precisely how straight lines would seem to us if a fixed polar grid was our natural way of seeing the world. The r^2 term in the expression $ds^2 = dr^2 + r^2\, d\theta^2$ is how this is captured mathematically. It is the lines of the co-ordinate grid which are curved and not the surface itself.

However, it is not different parametrizations of one surface with its intrinsic metric that is of most interest. More important is the idea of different surfaces with different metrics, of which the simplest examples are the three types of surface we have already encountered: those of constant curvature. On each of these a geometry of geodesics can be built up; on the plane it is familiar

Euclidean geometry. On the sphere it is spherical geometry, very like Euclidean but with the gross difference that all lines (i.e. great circles) meet not only once, but twice (this global difference arising from apparently similar local situations is typical of the differential approach to geometry). On the pseudo-sphere we obtain Lobachevskii–Bolyai geometry with the regrettable existence of various geodesics between points. For the first time we have a way of saying what the various geometries are without making any question-begging assumptions about parallels.

The profound suggestion of Riemann is that basic to geometry is the notion of position, and relations of position can be expressed by means of direction and distance. From these basic notions it should be possible to recapture all of classical geometry and to invent new geometries which might be of independent interest, for example in physics.

To show that the Lobachevskii–Bolyai geometry truly exists it is necessary to define a metric on a surface which allows geodesics to be asymptotic. Without engaging in the analysis we cannot really do this, but two suggestive processes can be demonstrated. One is to regard these grids as basic, where the a's are a pencil of parallel lines and the b's are a family of associated horocycles. It is a type of polar grid. Another way is to return to the canonical projection of Lobachevskii and use it to push the co-ordinate grid from the Euclidean plane (i.e. the bowl-like horosphere) onto the non-Euclidean plane below.

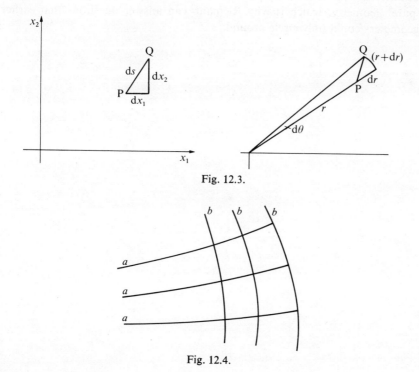

Fig. 12.3.

Fig. 12.4.

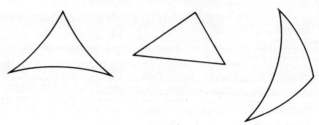

Fig. 12.5.

If either of these processes can be carried out, and they can, then they legitimize the formulae of Lobachevskii and Bolyai, which become formulae for geometry on Riemann's definition. Whereas before the analysis was empty—it is unclear what replacing k by ik is doing—it is now the introduction of a new metric. The analysis has merely to be consistent, which was known already, to complete the picture.

Most pleasingly we can now say what is curved or straight about the sides of the triangles in Fig. 12.5, and the answer, which is simple, but not obvious, is that in different geometries they are either straight (i.e. geodesics) or not depending on the surface and its associated metric on which they are drawn. The intuitive concepts of straight and curved are anchored in a discussion of basic notions, as is the concept of distance, in a way that they were not in old 'pure' geometry, which is why Riemann can answer questions that earlier geometers could only circle around.

13 Beltrami's ideas

Riemann's ideas spread only slowly. Beltrami seems independently to have reached a similar position over Lobachevskii's work, of which he published a suggested interpretation in 1868. Happily for our purposes it is much more intuitive than Riemann's, being descriptive in the classical sense. It also had the virtue of clarifying what is meant by the validity of a geometry.

Take one of the equivalent formulations of the parallel postulate, namely the one concerning the non-existence of similar triangles. A theorem to this effect met with the following reactions.

(1) There are similar triangles in reality, so the parallel postulate is proved.
(2) (in answer to (1)). Not at all, the theorem only asserts the equivalence of the existence of similar triangles and the parallel postulate, but this bizarre result still inclines us to reject alternative geometries.
(3) (in answer to (2)). Unfair; the mathematics cannot be challenged. However, we can experiment to determine which world we live in (since the non-existence of similar triangles implies a connection between angle sum and area).

In the transition from (2) to (3), a view held by Lobachevskii and Bolyai, the vexed question of mathematical truth arises. Mathematics has divided into two, a part with an empirical content and the rest. For this infinitude of other worlds what is the logical status of their descriptions? The passage where Lobachevskii exhibits the transition via absolute spherical trigonometry from Euclidean to non-Euclidean geometry suggests a way to approach the question: the idea of relative consistency. Beltrami's model first of all demonstrates that the new geometry is consistent if the old one is by showing that any inconsistency in it would appear as an inconsistency in the Euclidean geometry.

Let us return to the canonical projection map of Bolyai and Lobachevskii. The picture is of one surface resting upon another. On the lower surface straight lines satisfy the axioms of non-Euclidean geometry. The whole plane below is the image of a circular region on the upper surface, or, more strictly, of the interior of that region, for its boundary Ω is defined by lines asymptotic to the lower surface. If we draw this interior region we obtain a manageable picture of the whole non-Euclidean plane in which straight lines below correspond to parts of horocycles.

Imagine light streaming up from the table along the paths that form the pencil. In this way the whole table is cast as a picture onto part of the bowl, although distances are assessed differently on the two surfaces. Can we use this way of drawing a picture of the non-Euclidean plane on part of the

136 Beltrami's ideas

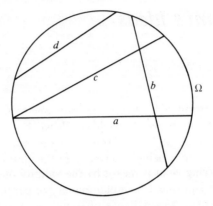

Fig. 13.1.

Euclidean one to tell us about non-Euclidean geometry? First of all, what are the properties of lines and triangles, for instance, in this model? (Such a picture is a model of non-Euclidean geometry.)

Lines a and b meet, whereas a and d do not meet. They do meet if you prolong them outside the circle, but that is just what we do not do because under the canonical map the points outside have no image on the plane. Nor are the points on Ω 'there', so the lines a and c do not meet in our model. They are parallel, whereas a and d are ultra-parallel. If we wished to follow Riemann we would introduce a metric on the interior of Ω which allowed us to recapture the familiar formulae, which is what Beltrami did, but let us instead, still following Beltrami, try to express our results in the language of classical geometry.

We need to have points and lines defined, and the axioms of Euclid except for the parallel postulate checked with these definitions. It is easy enough to check them all, and they turn out to depend in the new case upon their validity in the old situation. Suppose we wished to check that any two points lie on exactly one line: we draw the line through them exactly as in the Euclidean case and keep the segment of the line inside Ω. This constructs a line through the two points, and it is unique since, if there were two distinct lines removing Ω would exhibit two distinct line segments through two points in the Euclidean plane.

Rather than considering the interior of a circle with a cunning metric, can we not consider a surface drawn in three dimensions and on which there is a natural metric—just like a sphere? Indeed we can; it is the pseudosphere of Minding, which was also considered by Riemann. Beltrami also showed that his cunning metric in a disc was but a strange map of the pseudospherical world. Upon the pseudosphere distance is defined in terms of the lengths of geodesics. A triangle on the pseudosphere, made up of segments of geodesics, has sides and angles which we can measure and compare and which satisfy certain trigonometrical formulae.[1] The formulae are those of Taurinus, Lobachevskii and Bolyai, so the pseudosphere is indeed a world in which the new geometry is valid. However, and this is Beltrami's point, a simple dictionary allows us to translate statements about geodesics on a pseudosphere into statements about lines in a non-Euclidean geometry and vice versa. Any self-contradictory results in non-Euclidean geometry would then be interpretable as self-contradictory results about geometry on a surface in three dimensions and therefore as self-contradictions in the Euclidean geometry of surfaces and space. In this way the relative consistency of non-Euclidean and Euclidean geometry is established.

A region of the Euclidean plane is made to exhibit the descriptive features of a new geometry, and the consistency of the new geometry is seen to depend on the old. Such a procedure is known as creating a model of the axioms, and Beltrami's model of non-Euclidean geometry is historically the first example of this approach. To complete it, it would be necessary to define a metric on the interior of Ω which has the desired properties, and it will be rather bizarre. Chords in the disc must become geodesics and the non-Euclidean length of a chord must be infinite. Beltrami did this as well (see below). There is for now a more important philosophical point to be made.

It is often argued, somewhat incorrectly, that the discovery of new geometries was held back by specifically Kantian views of mathematical truth: that mathematics revealed *a priori* truths about the world. Since there is only one world, it would follow that there is only one true geometry of it. A more honest appraisal would recognize that mathematicians believed that the results of their work were truths and that this world was validly described in mathematics. However, on either view mathematics is anchored in reality, and the truth of it was frequently reinforced by reference to the nature of the world. As long as the hypothetical worlds were trivial and bizarre nobody needed to worry. As the descriptions of a non-Euclidean world grew they elicited at first precisely these feelings of rejection: in crisis mathematicians believed in the world they walked upon as the only one. So deep did this crisis go and so profound were the consequences of its resolution (we have scarcely begun to explore them here), that this view of mathematical truth itself could not survive. It is an implausible view that out of an infinitude of possible worlds of equal richness we should inhabit any one in particular (how would God know which one to select?). More precisely, how can mathematics reveal truths about this one specifically to the exclusion of the others? For, once it is admitted that mathematics describes all worlds with equal force, and this one quite impartially, what becomes of its truthfulness? We can no longer validate mathematics by experiment if it is to be equally valid in quite different physical worlds. It seems that for a time it was enough to settle the question of

---

[1] Beltrami could here rely on a paper by Codazzi (1857) in which it had been shown that the intrinsic trigonometry of the pseudosphere was hyperbolic. However, Codazzi had missed the connection with non-Euclidean geometry and was unaware of Riemann's work.

the truth of the new mathematics by exhibiting its consistency relative to the old, but this resolution is incomplete.

It is possible to worry about the truth of all of mathematics. It is not enough to make the validity of the new depend on the old if the old geometry is likewise under question. What Beltrami's model does is to establish a relative consistency between one bit of mathematics and another, which would indeed confer truth on the new bit if only it was known that the old was true. However, that is just what was not certain. How might the truth be established? One course is open if you believe in scientific truth, as the nineteenth century only too eagerly did. A patient scientific examination of reality could elaborate an acceptable bed-rock, and a network of relative consistency proofs could make mathematics true and meaningful. We would then inhabit a particular type of world (perhaps with $K = 0$, the Euclidean) and around us would be other worlds, just as true but not open to us. They would be mathematically valid, but non-existent. Mathematics would be the only glimpse we could ever have of them.

A second, more prudent, course is to establish the other branches of mathematics in a hierarchy, so that the Euclidean geometry rests on something more basic in some sense, and hope to find indubitable truths at the bottom of the pile. The obvious source is that perceived by the Greeks: logic. Perhaps mathematics can be shown to be relatively consistent with logic, and it does not seem to make sense to deny logic.

It must be stressed that the method of models is more creative than the classical method of deduction. A model is both a set of axioms (and deductions) and a link to another set of axioms. As in the example of geometry, the two different sets of axioms may, taken together, contradict each other. All a relative consistency proof asserts is that if the first set alone could lead to contradictory deductions, then so can the second set taken on its own. Either we accept the parallel postulate, or we deny it in some precise way by fixing a value for $K \neq 0$. We cannot and do not do both at once, but we do assert that if we fix a value $K \neq 0$ we shall only get into trouble if we would have done so by opting for $K = 0$ in the first place.

The first course makes mathematical truths empirical, i.e. it makes it an empirical question as to whether or not a mathematical statement describes the world. The second course, which looks for an unassailable basis for mathematics in a logical or other formal system, was attempted by several people, notably Russell and Hilbert. The curious and surprising fate of these attempts is well treated elsewhere (for example by Körner (1971)) and it is too long a story to tell here. It turns out that it cannot satisfactorily be done, at least not with the criteria for satisfaction proposed by original workers in the field. However, even if one turns aside from these fundamental questions—and one may indeed wonder (with, for example, Lakatos (1976) and Putnam (1974)) if mathematics has any need for foundations of the kind discussed above—there remains the fascinating question of the validity of particular kinds of mathematical argument. I suggested above that the peculiar power of the arguments of Bolyai and Lobachevskii was that they were rooted in analytic techniques rather than those of classical geometry. Now that a differential geometric formulation of geometry has been given, it will be interesting to look again at that analysis.

Lobachevskii himself seems to have made a connection of some sort between geometry and analysis. In Beltrami's language we could say that he was striving to exhibit a relative consistency between his new geometry and analysis by exhibiting a translation from the new geometry to formulae in analysis. Such a model is harder to establish and calls in any case for more clarity about fundamental notions than Lobachevskii possessed, but it would go something like this.

Each geometrical entity would become some analytic entity, for example, a variable or a function. Each geometrical property would become some analytic relationship (equality, inequality, identity) (see below for an example). It must then be shown that an inconsistency in the geometry (say, that parallel lines are unique after all) results in a contradiction in the analysis (say, that a strict inequality between two things must also be an equality). Lobachevskii did not do this, however. Although he did exhibit impressive formulae, this is not enough. The new geometry could be wrong all alone and still yield pretty formulae, since falsehoods can yield truths. All logic forbids is that truths should yield falsehoods (by their fruits ye may remain confused), so what he should have tried to do was to obtain falsehoods from his formulae or, better, prove that his formulae could never self-contradict by capturing all of the geometry in the analysis, say in a list of formulae, and then separately deriving those formulae entirely from basic analysis. However, this requires more knowledge than the mathematical community was to have for some time. It was not until the advent of David Hilbert (1898–9, see also Wolfe 1945) that a complete axiomatization of Euclid was achieved and then the whole real line was incorporated into geometry—which may yet turn out to be excessively greedy. Blumenthal (1961) remarks: 'Hilbert's postulational system is as consistent as the arithmetic of real numbers'.

The analytical model of Euclidean geometry

Cartesian or co-ordinate geometry is a model of Euclid's in a sense that Lobachevskii would have understood. To a point in the plane we associate a pair of real numbers x and y thus, (x, y). The plane is the set of all such pairs of real numbers (x, y). A line is described as the set of (x, y) such that, for some a and b,

$$\frac{x}{a}+\frac{y}{b}-1 = 0;$$

b/a is called the gradient of the line. Since two lines are parallel if and only if they never meet,

140 Beltrami's ideas

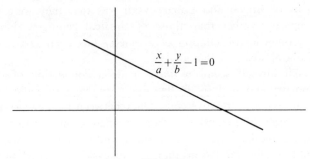

Fig. 13.2.

$$\frac{x}{a} + \frac{y}{b} - 1 = 0$$

and

$$\frac{x}{a'} + \frac{y}{b'} - 1 = 0$$

meet where

$$x = \frac{aa'(b-b')}{(a'b-ab')}$$

and

$$y = \frac{b'b(a-a')}{(b'a-ba')}$$

which point exists unless $a'b - ab' = 0$, i.e. unless $b/a = b'/a'$ and the lines have the same gradient. Therefore the line through P parallel to $x/a + y/b - 1 = 0$ is the unique line through P with gradient b/a.

If we follow Riemann we must define a metric

$$d\{(x_1, y_1), (x_2, y_2)\} = \{(x_1 - x_2)^2 + (y_1 - y_2)^2\}^{1/2}$$

and deduce that indeed $x/a + y/b - 1 = 0$ is the equation of a geodesic. We can also deduce the formula for the tangent of an angle between two lines, but we do not. Instead, observe that a different metric could well produce different geodesics (straight lines) and consequently different results concerning parallel lines.

Let us now consider the formulae in hyperbolic trigonometry that Lobachevskii established for hyperbolic geometry. He derived them on the *assumption* that a non-Euclidean geometry could be valid, but let us not make any assumption about that either way and begin simply with the formulae which are valid as pieces of analysis. They can now be interpreted as (i) a map from a Euclidean plane (the horosphere or the flat version, it does not matter which) to a surface which defines the shapes of the formerly Euclidean triangles, and which (ii) also defines a metric on that surface. That surface is, naturally, non-Euclidean. Therefore the formulae are indeed enough, but for Lobachevskii to have said so he would have had to add to his already brilliant record the conceptual discoveries of Beltrami. We may, however, imagine that he intuitively felt that something along these lines had to be true.

14 New models and old arguments

In the historical exposition of the subject the emphasis was on a geometry with many lines not meeting a fixed line. The other alternative is a geometry with no parallels, a possibility considered and disproved by Saccheri and Legendre amongst others. Yet all along there was the spherical geometry (of great circles on a sphere) in which there were indeed no parallels. With the reconstruction of geometry we can return to this contradiction and attempt to resolve it. Historically Riemann was the first to do so; the crucial idea occurs in the *Hypotheses*, III, 2 where he argued that lines may be unbounded without also being infinite. Lines might resemble circles on a sphere, and if we are to adopt this suggestion we might do so in the spirit of differential geometry. However, there is one trivial blemish on spherical geometry which may have consoled (or misled) workers before 1854: any two great circles on a sphere meet not just once but twice at diametrically opposite points.

The way out, suggested not by Riemann but later by Felix Klein[1], is to consider not the whole sphere but exactly half of it. We make a uniform selection of points on the sphere, selecting one from each pair of diametrically opposite ones so as to take, say, the top half (N.B. with exactly half the equator). We no longer have great circles, but instead halves, and each pair meet exactly once. In this way we obtain a picture of a geometry whose lines are halved great circles. To complete the programme we must define a metric on the surface, but this is easy. We can use the familiar surface distance which leads us to consider great circles in, for example, navigation. On a fixed sphere of radius R this is proportional to the angle subtended at the centre, and we shall use this angle as the measure of distance.

The trigonometry of the surface is the usual spherical trigonometry, and spherical geometry is in this way realized as the geometry of geodesics on a surface of constant positive curvature. It changes its name to projective or, more strictly, elliptic geometry, but never to Riemannian geometry, which name is reserved appropriately for the general study of manifolds with reasonable metrics.

The finitude of lines is its most novel feature; no line can be as long as, or longer than, πR (cf. p. 64). Once this is accepted the geometry is easily handled. For instance, what is the circle centre E, radius ρ? A point F on the circle must lie on an axis of the sphere inclined at an angle ρ to OE. Such a line lies on the cone with its vertex at O and vertical angle ρ, and when this is

---

[1] F. Klein, Über die so-genannte nicht Euklidische Geometrie, I, § 11, *Math. Ann.* **4** (*Ges. Math.*, Abt. I, 16) (1871). In differential geometric terms Klein considered geodesic projection.

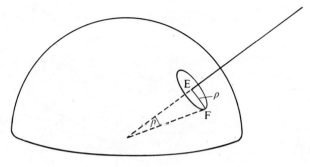

Fig. 14.1.

drawn it is seen to cut the sphere in a pleasingly circular shape. The angle between two lines in elliptic geometry is the angle made by their tangents in the familiar Euclidean sense.

The historically interesting thing is that this simple description of a non-Euclidean geometry was not made very much earlier. That it could not be was due to the inadequate grasp of fundamentals; only on a Riemannian programme could the arguments above be seen to be geometric and this simple model be understood as illustrative of anything comparable with Euclid's geometry of space. Lambert, Kant, and Taurinus rejected spherical geometry as a possible geometry for space on the grounds that in it two lines can enclose an area.

In order to revisit the Euclideans (Saccheri to Legendre) it is necessary to develop a better model of hyperbolic geometry than we have so far expounded. The drawback to Beltrami's model from our point of view is that it is visually far from suggestive. There is a deep reason for this, which was to escape mathematicians for another 50 years after Riemann. A metric for elliptic geometry was easily found once we could draw the whole surface isometrically in a Euclidean space (admittedly of dimension 3). No such procedure is available for hyperbolic geometry, as David Hilbert[2] showed in 1901. Minding's surface, the surface of rotation of constant negative curvature, can be drawn in various ways in three-dimensional space, but never without singularities, i.e. never without there being a region which the geodesics cannot penetrate and consequently the two-dimensional manifold (or surface) appears to have a boundary. This is an accident arising from arbitrarily trying to draw the surface in three dimensions, in the way that two-dimensional pictures of knots have self-intersections. Consequently a natural metric which can also readily be visualized cannot be defined.

Something can be done, however. We keep the idea of modelling hyperbolic geometry on a circular region of Euclidean space, and so the metric will

---

[2] D. Hilbert, *Über Flächen von konstanter Gauss'scher Krümmung* (transl. *Am. Math. Soc.*), pp. 86–99 (1901); *Grundlagen der Geometrie*, pp. 162–75 (1898–9).

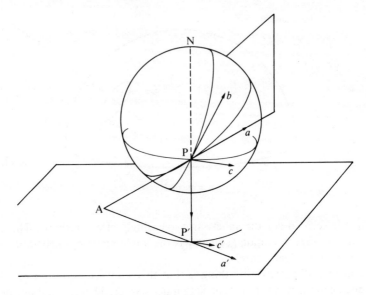

Fig. 14.2.

look pretty odd since hyperbolic lines are infinite in length. However, we might be able to define angles in a way which looks better. Such a way was discovered by H. Poincaré in 1881 and relies upon the method of stereographic projection; we give a version here which is immediately equivalent to his (see his *Oeuvres*, II, p. 1).

Imagine a sphere resting upon a plane with the north pole uppermost. The sphere can be projected, or mapped, onto the plane by sending each point on the sphere along the line from the north pole through it until it meets the plane. Only the north pole is not mapped anywhere according to this rule, but we need only consider a limited region of the sphere anyway. The essential feature of this mapping is that it preserves angles and circles (see Fig. 14.2).

An angle between two great circles on a sphere is the angle between their tangents at that point, so we reduce Fig. 14.2 to Fig. 14.3. Take Fig. 14.3(a) first. The plane defined by NP and a, the tangent at P, meets the ground in a', and a and a' meet at A, say. Denote the angle between a and c by α and the angle between the projections a' and c' by α'. It is enough to show $\alpha = \alpha'$, for then a similar argument shows that in the other figure $\beta = \beta'$ and then $\alpha - \beta = \alpha' - \beta'$ which is the desired result. To show that $\alpha = \alpha'$ note first that all latitudes go to concentric circles and the angle between NP and c is therefore the same as the angle between NP$'$ and c' (both are right angles). Therefore, looking at the angle in the plane of NP and c, we have, for reasons of symmetry, AP = AP$'$, from which it follows that $2R - \alpha = 2R - \alpha'$ and so $\alpha = \alpha'$ and the result is proved.

We can now derive Poincaré's model of the hyperbolic plane from Beltrami's flat model by two maps. In a circular region m of the plane, having radius r, draw Beltrami's model (Fig. 14.4(a)). Stand a sphere, also of radius r, on the plane with its centre vertically above the centre of m (Fig. 14.4(b)). By vertical parallel projection (not stereographic) project m onto the sphere. The boundary of m will lift to the equator and m itself will be mapped onto the southern hemisphere (Fig. 14.4(c)). Now project the southern hemisphere back onto the plane by stereographic projection from the north pole. It will cover a circular region M of radius R (Fig. 14.4(d)). It is this region that we shall take as our model of the hyperbolic plane. What has happened to a non-Euclidean straight line in m under this process? Originally a chord of m, it rises vertically under the first map to become an arc of a circle on the sphere (not usually a great circle) which meets the equator at right angles (Fig. 14.4(e)). Under stereographic projection this circle is mapped into a circle meeting the boundary of M at right angles. Therefore hyperbolic straight lines are modelled by circles perpendicular to a fixed circle. As it turns out, when the metric is put into the model the hyperbolic angles are identical with the Euclidean angles they seem to be (Fig. 14.4(f)).

The Poincaré model can be built up from first principles if you do not want to go from the Beltrami model. Let us call a hyperbolic straight line the segment of any circle contained within a fixed circle S and perpendicular to it. We shall show that there is a line between any two points. Take A and B in S, and look at all the Euclidean circles through A and B. Some do not meet S,

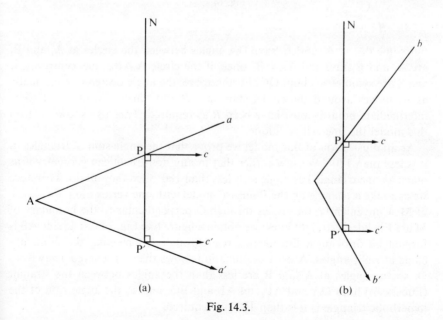

Fig. 14.3.

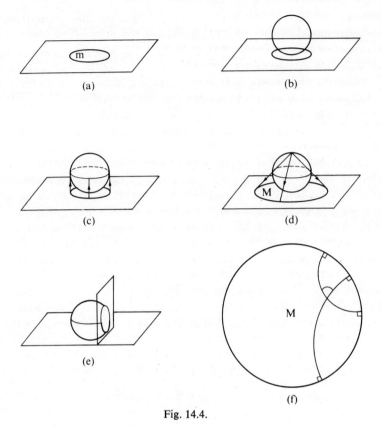

Fig. 14.4.

but some do, at A_i and B_i say. The angles between the circles at A_i and B_i are $\hat{A}_i$ and $\hat{B}_i$ say, and $\hat{A}_i = \hat{B}_i$ since, if the circle A_iABB_i has centre C_i, it and S are symmetric about OC. Furthermore, the angle decreases continually as A_i moves around the circle, starts at $2R$ and finishes at O, so at some intermediate point it must have been R as required. This also shows that on this model the line AB is unique.

As an illustration of this model we prove that the angle sum of triangles in it is less than 180°. We saw earlier that in any non-Euclidean geometry if as much as one triangle has angle sum less than 180° then they all do. Therefore we can take a triangle on the Poincaré model with one vertex at O, the centre of M. Conveniently, the circles through O perpendicular to the boundary of M are Euclidean straight lines; we obtain a figure like Fig. 14.6. The side AB is formed by drawing a Euclidean circle through AB meeting the boundary circle at right angles. A brief examination shows that it is curved away from O, so the angles at A and B are less than the angles between the straight (Euclidean) lines OA and AB and AB and BO, and so the angle sum of the hyperbolic triangle is less than 180° as required.

New models and old arguments 147

Distance on this model remains hard to visualize. Parallel lines are once again those which meet on the boundary, which is infinitely far away. They are asymptotic, and non-meeting lines diverge. The best that can be done is to imagine a severely non-linear scale of distance, so that points the same Euclidean distance apart steadily diverge as you move towards the rim (see p. 201).

With this model of hyperbolic geometry and the Riemann–Klein model of the elliptic plane, we can now revisit earlier ground and see how people made the mistakes they did.

Classical geometry revisited

The earliest attempts on the postulate aimed to show that the curve c everywhere equidistant from a straight line l was itself straight. This cannot be the case. In elliptic geometry the equidistant curve is a circle, and in hyperbolic geometry it is an arc of a (Euclidean) circle meeting the boundary at an angle which is not a right angle. Wallis's construction of similar triangles is more interestingly flawed. In the elliptic case lines a through B and b through A must meet anyway, at C say. A line through B_1 such that $\beta_1 = \beta$ meets b at C_1 but the top angle has changed, and indeed is necessarily smaller than the angle

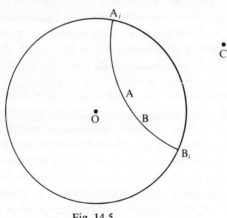

Fig. 14.5.

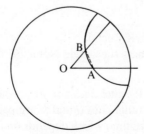

Fig. 14.6.

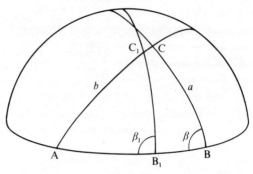

Fig. 14.7.

at C. Hyperbolically, we have a picture like Fig. 14.7, in which we may suppose that AB lies on a line through the 'centre' of the region. Again, the top angle of the triangle in Wallis's construction varies with the area of the triangle and so similar triangles cannot be found.

With Gerolamo Saccheri we recommence non-Euclidean geometry proper. Freed of his vain pursuit of contradictions we may restate his work this way. In elliptic geometry the fundamental figure is the quadrilateral ABCD, which has right angles at A and B and AD = BC. We can conveniently represent this figure on the sphere with AB part of the equator and AD, BC two longitude lines. DC, to be straight, will be a segment of a great circle, and we see that $\hat{D} = \hat{C} \geq 90°$. Furthermore $AB \geq DC$, since the lines of longitude ultimately meet, and all quadrilaterals share the obtuse angle property with this one.

Saccheri's refutation of such a non-Euclidean geometry was based upon the following result: given a line perpendicular to a fixed line it will meet any line oblique to the fixed line. From this he deduced that any two lines meeting a fixed line and making with it angles whose sum is less than $2R$ must themselves meet. This is the original form of Euclid's postulate. In elliptic geometry we notice, not that this result is false, but that it is trivially true.

How then is Saccheri's construction avoided? In elliptic geometry l meets l' even if $\alpha + \beta = 2R$, which result Saccheri knew since it can be obtained by a trivial modification of his argument above on insisting on the HOA. Saccheri's argument was as follows.

(1) Assume the HOA.
(2) Deduce the postulate of Euclid.
(3) Deduce from (2) that the angle sum of a triangle is $2R$, which contradicts (1).

Step (3) runs like this. Parallel lines are lines which never meet, and Euclid has a construction for parallel lines (l and l' are parallel if they make supplementary angles with a third line) which depends on the proposition that in any triangle the sum of any two angles is less than $2R$ (*Elements*, I, 17). Since

$\alpha + \beta = 2R$ they cannot be angles in the same triangle. How do we know that the sum of two angles in a triangle is less than $2R$? Because the exterior angle is greater than either interior angle (*Elements*, I, 16). However, in elliptic geometry even this result is false, so how did Euclid prove it? By means of the following construction. In triangle ABC take the exterior angle at C and assume, if necessary by relabelling, that $\hat{B} \geqslant \hat{A}$. Let D be the mid-point of BC, and join AD and produce it to E so that $AD = DE$ (a). Join EC, and observe that the triangles BDA and CDE are congruent, so in particular $\hat{B} = A\hat{B}C = D\hat{C}E$. However, $D\hat{C}E < D\hat{C}Y = B\hat{C}Y$, so the exterior angle at C has been shown to exceed $\hat{B}$, and $\hat{B} \geqslant \hat{A}$, and the result is proved.

From (a) onwards our statements are valid in elliptic geometry. It is the construction of AE, the double of AD, which cannot necessarily be made. An attempt to do this is shown in Fig. 14.10. Notice that in this case the exterior angle at C is less than either of the interior angles at A or B.

We see finally that Euclid allows the indefinite replication of the lengths of segments along a line. The length of a line is infinite in his geometry. By contrast in elliptic geometry it is strictly finite. It is sometimes argued that the infinite length of lines was implicitly used by Saccheri in making his construction (p. 57), i.e. in step (2). This is so, but he need not have made it there; it is true that the construction of a perpendicular to one line and meeting another can always be made in elliptic geometry. It is indeed the case that lines l and l' inclined to m as described do meet; all lines meet in elliptic

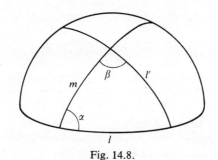

Fig. 14.8.

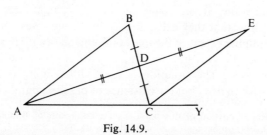

Fig. 14.9.

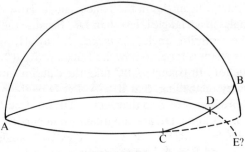

Fig. 14.10.

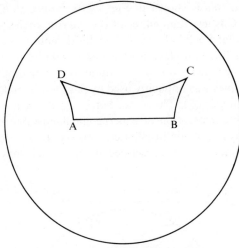

Fig. 14.11.

geometry and there are no parallels. Saccheri's fundamental use of the infinite length of the line is in deriving (3), when he appeals to those theorems of Euclid which can be established without reference to parallels at all.[3]

What Saccheri discovered was a deep connection between length, angles, and parallelism quite unsuspected by Euclid. The existence of these connections was to force a thorough re-examination of all of Euclid's premises, culminating in the work of Hilbert. An equivalent way of stating again what we have just said is this. If you define parallels as straight lines which do not meet you must show that they exist, and to do this involves making assumptions about the infinite extent of the line (as Riemann saw). Therefore denying the existence of parallels involves denying at least one proposition on the way to proving they exist. The situation is quite different if, as in hyperbolic geometry, you wish to deny not the existence of parallels but their uniqueness.

---

[3] For an excellent discussion of this point see the comments of Bonola (1912 (reprint 1955), pp. 30, 120, 144) on Saccheri and on Dehn's work.

New models and old arguments   151

In hyperbolic geometry Saccheri's quadrilateral is as shown in Fig. 14.11, by sliding it into a symmetric position. Once again the elementary equalities and inequalities of this figure are evident, e.g. AB < DC. However, we cannot now necessarily construct a line l' perpendicular to l and meeting m, a line oblique to l. Equal increments along m, in the hyperbolic metric, seem to crowd together in the Poincaré picture. The 'last' perpendicular from m to l cannot be drawn, but none can lie to the right of n. In this way the key step in the HOA argument fails to pass over the HAA case, as Saccheri rightly concluded (Fig. 14.12).

Saccheri next considered asymptotic straight lines, which are, of course, parallel in the new scheme. In this way he was led to the familiar figure for parallels, which in the new model is as shown in Fig. 14.13.

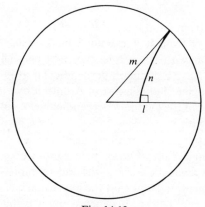

Fig. 14.12.

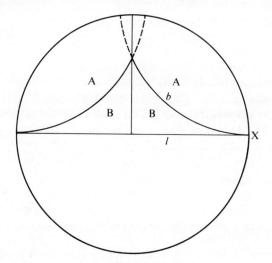

Fig. 14.13.

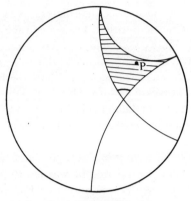

Fig. 14.14.

Lines in A diverge from *l* and lines in B meet *l*. The left and right parallels through P to *l* diverge from it if extended backwards. A line in A has a common perpendicular with *l*. Let *b* be the right parallel. He concluded by asserting that *l* and *b* have a common perpendicular at infinity. Meaningless as this is, you can imagine sliding a line in A until it overlaps *b*, watching all the time the behaviour of the common perpendicular. It collapses to X, taking Saccheri's argument with it.

The final classical approach we shall reconsider is one due to Legendre. His argument that the angle sum of a triangle cannot exceed $2R$ (p. 70) self-evidently requires the replication of lengths and so implies the infiniteness of lines. His other argument, to show that the angle sum cannot be less than $2R$ (p. 71) is more amusingly flawed. It depends upon being able to make the following construction: from any point within an angle draw a straight line cutting both arms of that angle. On Poincaré's model, however, the picture is as shown in Fig. 14.14, and Legendre's construction can only be made when P is within the shaded part of the angle, bounded on the right by the line asymptotic to the sides of the angle.

Interestingly enough, the thirteenth century Arabian mathematician Attin Eddin al Abhari (flourishing 1264) had made the same mistake, although not without investigating it first. He argued that if an angle BÂC is bisected by a line LA then the perpendicular to LA through any point on it will meet both BA and AC. The argument was independently resurrected by T. Cullavin and published in Cayley's *Journal of pure and applied mathematics* as late as 1894, together with a belated, and incorrect, refutation of it by Cayley himself.[4]

Finally, we might look for a pleasing representation of the horocycle on Poincaré's model. Recall that it is the locus of points corresponding to a given one with respect to a pencil of parallel lines and that it meets each line of the

---

[4] I am indebted to Dr. Mehdi of Birkbeck College, London, for drawing my attention to the material in this paragraph.

pencil at right angles. A pencil of lines can easily be drawn, converging on X say, and notice that the diameter through X is a member of the pencil. All other members of the pencil, because they must be perpendicular to the boundary circle, are tangent to the diameter at X. The horocycle must be perpendicular to d (Fig. 14.15). Consider a (Euclidean, not hyperbolic) circle C, tangent to the boundary circle at X and having part of d as its diameter. This circle turns out to meet every line of the pencil at right angles, and so is a horocycle for the following reason. Let a be any member of the pencil. The angles at which C meets a are equal, and they meet at X at right angles by construction. Therefore, at their point of intersection in the disc they also meet at right angles and C must be a horocycle. Recall that X is not a real point in the model. This description of the horocycle can be found in Beltrami's *Saggio* (1868).

Exercises

14.1 Show that the Poincaré model satisfies all the axioms of Euclid's *Elements* except the parallel axiom.

14.2 Draw the canonical projection map of Lobachevskii and Bolyai in two dimensions.

14.3 When the disc is rotated about a diameter it sweeps out a sphere. The interior of this sphere becomes a model for a non-Euclidean space of three dimensions and a horocycle perpendicular to that diameter sweeps out a horosphere. The non-Euclidean metric of the ambient three-dimensional space induces a Euclidean metric on the horosphere. Use this picture to

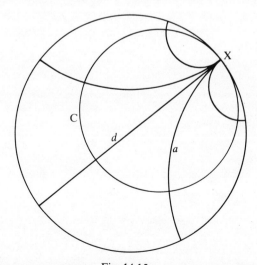

Fig. 14.15.

reconstruct the original arguments of Lobachevskii and Bolyai. It is surprising and amusing that, although we cannot satisfactorily obtain a surface in Euclidean space on which a non-Euclidean geometry holds, non-Euclidean beings can very easily obtain a copy of our world.

14.4 Show that non-Euclidean circles are precisely those Euclidean circles lying entirely in the disc, but that their Euclidean and non-Euclidean centres only coincide if both are at the centre of the disc.

15 Resumé

The publications of Riemann and of Beltrami in 1867 and 1868 decisively transformed the study of non-Euclidean geometry and thereby mark a convenient break in the story. Most historians, following Bonola's example, have indeed ended their accounts with a discussion of those influential papers. Therefore it is natural to provide in this chapter a brief summary of the developments that have so far passed under review, and to link it with criticisms and modifications of the accounts that have been given by other writers.

Bonola (1912 (reprinted 1955)) divided the post-1600 history into four periods: one for forerunners, one for Gauss, Schweikart, and Taurinus, one for Bolyai and Lobachevskii, and one on later developments. Coolidge's account (1940 (reprinted 1963)) likewise distinguished the period up to Bolyai and Lobachevskii from what he called the modern study of the problem by Beltrami, Riemann, and others. Kline devotes Chapter 36 of his *Mathematical thought from ancient to modern times* (1972) to the work done before 1840–50, and Chapter 38 to a further discussion of non-Euclidean geometry in the various contexts of projective geometry, metric geometry, models and consistency, and the question of reality. The intermediate chapter is devoted to a discussion of the differential geometry of Gauss and Riemann. The extent of agreement between these historical accounts greatly exceeds the extent of disagreement, and it will be convenient to regard them as providing what I may call the standard account. As such, the standard account gives a very good treatment of certain aspects of the subject. The division into periods coincides not only with the evident chronological divisions but with differences in mathematical methods: in the eighteenth century Saccheri and Lambert used classical geometry; in the early nineteenth century Bolyai and Lobachevskii used analysis; in the mid-nineteenth century Riemann and Beltrami turned to the techniques of differential geometry. It is also the case that before, roughly, 1800, mathematicians hoped to show that Euclidean geometry was the only possible geometry of space, whereas afterwards they sought to establish the possible validity of other geometries.

However, the standard account is open to several criticisms.

(1) It is well over 100 years from Saccheri's *Euclides Vindicatus* to Beltrami's *Saggio*, but the standard account does not explain why the development took so long.
(2) It does not explain, or adequately discuss, the choice of methods used at various times but subordinates it to a compilation of results. In Bonola's account, for instance, analytic methods appear unheralded in the discussion of Gauss, Schweikart, and Taurinus. This results in a failure to appreciate what it was that the mathematicians were intending to do.

156 *Resumé*

(3) The exact nature of the accomplishment of J. Bolyai and Lobachevskii is not fully discussed. For, once it is admitted that it is not logically conclusive it must be asked why it has been found so compelling. This is the problem which makes it very hard to say who invented, or discovered, non-Euclidean geometry.

(4) Finally, it does not explain why it is that spherical geometry, well known throughout the entire period, did not at once settle the question, yet this geometry is now given almost immediately in modern textbooks as an example of a non-Euclidean geometry. This criticism bears with particular weight upon that part of the standard account which sees the problem as one in foundations: the existence or non-existence of a geometry differing from Euclid's in respect of the parallel postulate. If spherical geometry is such a good example of a geometry based on the HOA then the complete confidence of everyone from 1733 to 1854 in the impossibility of such a geometry is a historical problem which the standard account does not resolve. To say that spherical geometry is irrelevant because it flouts the 'Archimedean postulate' (that lines can be indefinitely extended) is no answer, because no one argued that way at the time.

It will be seen that these criticisms highlight one pervasive failing in the standard account; its tendency to see the subject as a prolonged attempt to answer one question (is the parallel postulate necessarily true?). The problem is essentially, therefore, regarded as one in foundations, and is resolved when the (negative) answer is given. Indeed, the standard account frequently ends with references to the logical independence of the postulate from the rest of geometry (Bonola, Chap. V, § 94, Appendix V; Coolidge, pp. 84–8; Kline, Chap. 38, §4, Chap. 42). This is partly the result of a historical accident. During the first half of this century mathematicians expended considerable effort in providing axiomatic foundations for various parts of their subject, and undoubtedly one of the reasons for this was the discovery of non-Euclidean geometry. However, much of mathematics was in need of the clarity that an axiomatization can bring, and if anything it was projective geometry (Freudenthal 1962) that stood in greater need of an axiomatization to emancipate it fully from Euclidean and metric concepts. The coincidence of the problem of non-Euclidean geometry from the standpoint of, say, Proclus to Saccheri and, again to take examples, Pasch and Hilbert at the turn of this century diverted historians away from other, perhaps more important, matters. In general historians of mathematics have been reluctant to discuss the changing nature of mathematical problems, preferring to produce a linear compilation of results.

The criticisms of the standard account can, I suggest, only be met by an account which dwells more on the mathematical methods and intentions of the actors in the historical drama, such as this book may provide and which I shall now summarize briefly.

Saccheri and Lambert virtually exhausted the classical approach to the study of non-Euclidean geometry, and it was only with the introduction of analytic techniques that a new start could be made. The hyperbolic trigonometric functions came in around 1760, and although Lambert used them elsewhere in his work he missed the connection with non-Euclidean geometry. When the connection was made—by Taurinus and Gauss—the result was a reformulation of the problem. For the first time a language was available which was flexible enough to discuss the problem in other than Euclidean terms. Taurinus, at least, regarded the formulae as defining some kind of a geometry, but not a plane geometry; he was unable to think of plane geometry in other than Euclidean terms (i.e. points, lines, and planes). Gauss's views, as I have discussed, are much harder to establish, for the modern attitude to geometry is implicit in his magnificent study of surfaces but he seems never to have drawn the conclusions for non-Euclidean geometry explicitly in his work.

Simultaneously with Gauss, but independently of him, J. Bolyai and Lobachevskii drew the radical conclusion from the trigonometric formulae that a new geometry was possible for space. They surpassed Taurinus not just because of the boldness of their conclusion but because the transition from spherical to hyperbolic trigonometric formulae was no longer formal but a clearly understandable mathematical transformation. Taurinus had only been able to replace real by imaginary quantities, much as Lambert had invoked the imaginary sphere. The new approach gave a much greater insight into the formulae that obtained, although the mysterious constant k still lacked explanation.

As has often been remarked, Bolyai and Lobachevskii nonetheless rested their geometry on formulae derived from the assumption that a non-Euclidean definition of parallels was possible; they did not conclusively establish that it was. This weakness in their argument may have contributed to the contemporary hesitance in accepting their work. None the less, by working directly with non-Euclidean three-space they broke with the tradition of seeking a non-Euclidean plane somehow embedded in a Euclidean three-space, and so implicitly sought an intrinsic non-Euclidean geometry.

The next developments were those in differential geometry. Gauss's profound conception of geometry as intrinsic to a surface took time to sink in, and the first papers on the surface of constant negative curvature (Minding 1839, 1840; Codazzi 1857; see p. 137) only established that the intrinsic trigonometry of the surface was given by the hyperbolic formulae and did not make the connection with non-Euclidean geometry. Without a thorough appreciation of the intrinsic nature of geometry it is impossible to run the arguments of Bolyai and Lobachevskii backwards and derive a geometry from the formulae, and it does not seem that either man was able to express his arguments in this way. The first to do that was Riemann. He and, independently, Beltrami were able to start from a discussion of an intrinsic metric and obtain a trigonometric description of various geometries, thereby rescuing the earlier arguments from their inconclusiveness and simultaneously explaining k as a curvature. Geometry was decisively reformulated in local terms rather than global ones: 'line' and 'plane' were defined in terms of geodesics and curvatures.

Two dramatic changes in the 'problem of parallels' may be isolated. One is marked by the introduction of analysis: the problem is transformed into the study of various formulae. The second is the recognition of the intrinsic nature of geometry and specifically curvature: this enabled a truly geometric grounding to be given to the trigonometry, which might otherwise have described nothing. Without an intrinsic formulation of differential geometry the study remains different from the study of space (which alone seems to have a natural geometry). The lack of such a concept in the work of Monge, the distinguished French differential geometer active around 1800, may have significantly limited the French appreciation of the possibility of non-Euclidean geometry.

All of these considerations bear on the problem of spherical geometry. It is clear that the simultaneous existence, up to 1854, of a valid spherical geometry with a valid refutation of any geometry based on the HOA means that the problem was not a problem solely of foundations. It was rather a problem about the nature of space and geometry; Saccheri, for instance, spoke of the 'nature of the straight line'. Spherical geometry can exist alongside a refutation of the HOA because spherical geometry treats of curved lines on curved surfaces and not of straight lines. It is only when the distinction between curved and flat lost its force that the refutation of the HOA lost its validity, and the finitude of lines in spherical geometry was emphasized.

It was also only after 1868 that an appreciation of the connection of geometric congruence and rigid-body motion was developed, largely by Helmholtz and Klein. That study goes well beyond the limits of this book, for it is the origin of the modern study of Lie groups, but it can be said that a study of those groups which correspond to rigid-body motions (in a certain sense) is also a study of a geometry, and Lie's classification of those groups in low dimensions showed that amongst homogeneous geometries the only surfaces are those of constant curvature that we have already encountered. Other aspects of the response of the mathematical and scientific community after 1868 are also interesting but must be omitted for reasons of space (see, however, the articles by Richards (1977, 1978) and Toth (1967, 1977), and, for a discussion of the so-called Helmholtz–Lie space problem, see Lie (1880).

Part 3

16 Non-Euclidean mechanics

But there's this that has to be said: if God really exists and he really has created the world, then, as we all know, he created it in accordance with the Euclidean Geometry, and he created the human mind with the conception of only three dimensions of space. And yet there have been and there still are mathematicians and philosophers, some of them indeed men of extraordinary genius, who doubt whether the whole universe, or, to put it more widely, all existence, was created only according to Euclidean geometry and they even dare to dream that two parallel lines which, according to Euclid, can never meet on earth, may meet somewhere in infinity. I, my dear chap, have come to the conclusion that if I can't understand even that, then how can I be expected to understand about God?

Thus Ivan in a long speech to his brother Alyosha shortly before the introduction of the Grand Inquisitor, in *The Brothers Karamazov*, which Dostoevsky wrote between 1878 and 1880.[1]

The impact of the discovery of non-Euclidean geometries was to undermine the traditional confidence in mathematics and physics. Dostoevsky's metaphor of God's Kingdom lying as far beyond our ken as the new geometries lay beyond Ivan's is a startlingly conservative interpretation, for it emphasizes the existence of that Kingdom. As Ivan goes on to say:

Please understand, it is not God that I do not accept, but the world he has created. I do not accept God's world, and I refuse to accept it. Let me put it another way: I'm convinced like a child that the wounds will heal and their traces fade away, that all the offensive and comical spectacle of human contradictions will vanish like a pitiful mirage, like a horrible and odious invention of the feeble and infinitely puny Euclidean mind of man ... Let even the parallel lines meet and let me see them meet, myself—I shall see and I shall say that they have met, but I still won't accept it.

For those of us condemned to work with our puny minds, the escape into non-Euclidean geometry has seemingly implied the opposite choice. Once it was possible to think of new worlds, the certainty we had in the old ones began to evaporate. Just as a rich, if unfamiliar, geometry holds in those worlds, so does a rich new physics. With only slight modifications the nineteenth century mathematical physics was shown to hold as well there as in Euclidean space.[2] It is perhaps possible to explain why this result is at least plausible. What Newton successfully formulated was a connection between force and change in velocity. In order for a body to remain either at rest or in a state of uniform velocity no forces are necessary. For an object sliding on a desk to remain doing so no force is necessary if the desk is smooth, but

---

[1] Dostoevsky, *The Brothers Karamazov*, transl. D. Magarshack, Penguin Books, Harmondsworth, 1958.

[2] The earliest demonstration I know is that of Lipschitz in the *Journal für Mathematik* (see particularly Vol. 72 (1870) and Vol. 74 (1872)).

if it is rough it applies a force to the object (friction) which slows it down. Consider then a triangle sliding upon a surface, which we shall take as a two-dimensional analogue of three-dimensional space. If the three corners and edges of the triangle remain at rest relative to each other, it will be called a rigid body; we can imagine it sliding over our surface. Being a triangle it can exist on either a plane, a sphere, or a pseudosphere, and it can slide freely upon any of these surfaces. Nowhere does it buckle. However, if it slides on a pear-shaped surface the points must move relative to each other to accommodate the triangle to the varying curvature of this surface. This would suggest a 'force', varying across the surface, which distorts the triangle, but the surfaces of constant curvature are free of such intrinsic 'forces' which interfere with constant motion along the surface. Therefore we can conclude that for them, but for no others, an object decently apart from all other objects will not change its state (of rest, or uniform motion in a straight line).

Once the new worlds were successfully proclaimed they seemed both to liberate the human mind and to separate man from the old unities. Although even Riemann still considered space as three dimensional, that conception too had to be set aside before the new ideas could fully match the old in richness and significance. In this concluding part of the book I propose to give some hint of the development of this new geometry of space as it is found in the special and general theories of relativity. Necessarily such an account is selective, both on historical and mathematical grounds, and I am aware of some of the risks involved. Notably, I should not wish to suggest that non-Euclidean geometry itself was a decisive influence on the research in physics. Einstein was later to pay generous tribute to Riemann's work, and his 'solitary and uncomprehended' genius, which provided the detailed mathematical formalism for his general theory, but it does not seem to have led his thought so much as given it language. Poincaré, who abstracted from non-Euclidean geometry ideas of immense importance in the mathematical theory of complex functions, never reinterpreted physics along fully relativistic lines. The influence, then, was indirect. In mathematics too it would be absurd to say that non-Euclidean geometry alone shaped the future. The ideas of Riemann himself, for instance, were of vastly greater application than the new geometries in themselves. If we agree that the work of Bolyai and Lobachevskii stimulated work in two areas—foundations and geometry itself—which were also being developed in other ways and for other reasons, we should not be too far wrong.[3] With that caution then, let us go on with the story.

[3] Foundational questions in geometry go back a long way. In the nineteenth century there is for instance Bolzano's *Elementargeometrie* of 1804. This current merges with the rigorization of analysis which characterizes the whole century and leads, for example, to mathematical logic. Geometry itself, notably projective geometry, was independently studied at first, for example, by Poncelet, Chasles, Steiner, von Staudt, and Cayley.

17 The question of absolute space

Newtonian space

The counterpart to Euclidean space in mathematics is Newtonian space in physics. It can be thought of as an enormous stage, across which pass the events that make up the universe: the enduring stars, the brief particles, ourselves. Inside this box everything has its position, its path, and its time, and the business of the scientist is to give a rational account of it all. The rise of science in its modern form is associated with a view that one should attend to observables—positions, paths, and times—rather than to innate properties, tendencies, or essential natures. Formerly circular motion had required no explanation; it was natural for the stars and planets to move on circles, and natural for round objects to rotate. The tension between natural motion and observed motion was resolved in planetary astronomy by adding extra circles, and not until Kepler was a 'non-circular' orbit suggested for a planet (Mars). Here the non-circularity means an orbit not to be described as a combination of circles, and which therefore provokes the question: what moves the planet? The first good answer was propounded by Newton with his theory of gravitation. Two bodies attract each other with a force, the strength of which depends upon their separate masses and the square of their distance apart,

$$F \propto M_1 M_1 / r^2$$

symbolically, and this force restrains their inertial tendency to fly apart. However, the nature of this gravitation is not discussed. Newton would very much liked to have explained it in order to see better how the cosmos obeyed the will of God. He wrote, in the first manuscript of the *General Scholium*, January 1712–13:

I have not yet disclosed the cause of gravity, nor have I undertaken to explain it, since I do not understand it from the phenomena. (Hall and Hall, 1962, p. 352.)

Nor could he explain the nature of space and time, which he called God's 'sensory', and nor could he give a convincing proof of God's dominion by showing that there could be no other final cause for demonstrated forces and motions. Newton knew that God was the cause of gravity, but since he was unable to demonstrate this he left it out of his *Mathematical principles of natural philosophy* (Hall and Hall 1962, p. 213; this discussion follows Holton 1973, p. 52).

Explanations as to *why* objects do as they do fell away as science advanced, and were replaced by descriptions as to *how* they do it. The regularities were given mathematical expression, and Nature was subsumed under 'laws' governing the observables. On the 200th anniversary of his death Einstein

(1927) observed of Newton with deep admiration that 'he was better aware of the weaknesses inherent in his intellectual edifice than the generations of learned scientists which followed him', and, elsewhere, that Newton had the courage to go ahead without a full understanding of the concepts of space and time and to do the doable, to describe the laws of acceleration.

There is an immediate connection between Newtonian space and Euclidean geometry of course, which Newton expressed this way in the Preface to his *Principia*

'[Geometry itself is] founded in mechanical practice, and is nothing but that part of universal mechanics which accurately proposes and demonstrates the art of measuring.'

The uniqueness of space, since God surely has only one sensorium or stage in which to act out his thoughts, confers upon it the logically possible properties assigned to it in Euclid's geometry. It was in this way that Kant regarded the impossibility of two lines enclosing an area as a property of space and not of lines. The nature of space impresses itself upon the observables, in this case the paths of two rays of light. However, we must distinguish two aspects of space which we might call the abstract and the operational. Among the properties of the abstract are that it is (i) homogeneous, i.e. any one point resembles any other, and (ii) isotropic, i.e. it has no preferred direction. Of course, physical space is neither, but that is because of the objects in it. The stage itself is bare. (It remains bare in non-Euclidean geometry also.)

Amongst the operational aspects are the numerical and quantitative accounts we give of motion. Here Newton was an avowed relativist. God may well see objects as they are, but we can only work with them as we see them.

Newton concluded the first section of the *Principia*:

But how we are to obtain the true motions from their causes, effects and apparent differences, and the converse, shall be explained more at large in the following treatise. For it was to this end that I composed it.

Relative motion

Let us take two different observers, Jean and Jo. They can stand in different places, face in different directions, and move with respect to each other. If they do so they will give different descriptions of the same events, but it might turn out that a simple transformation can be made from one description to the other which reveals their essential equivalence. (I am reminded of the man who passed me once in the street endlessly repeating: the sun went down behind the hill, the hill came up in front of the sun. . . .)

Equivalence here means with respect to the physical properties entering the description. Position itself is not held to be relevant to physics—the laws are the same everywhere, by (i) above. Relative position matters, but Jean and Jo will agree on the relative position of two objects even if they give them different co-ordinates. Orientation is not important as (ii) above asserts, so it does not matter if Jean and Jo face in different directions.

Does it matter if Jean and Jo move? As Newton expressed them, the laws of physics treat of changes in momentum or velocity, of acceleration and impact. They are not about steady constant velocity, for a body at rest or in uniform velocity along a straight line will remain so unless acted upon by a force. This suggests that the descriptions given by Jean and Jo should be essentially the same even if they are in a state of uniform motion relative to one another, i.e. motion along a line without spinning since rotation involves a change of direction and so of velocity. However, with the restriction that they may differ by moving with constant relative velocity, their qualitative descriptions should agree and their numerical descriptions of events should differ in some simple way involving no physical concepts for their explanation. No forces are apparent to Jean but not to Jo; no forces are needed to explain the transformation from Jo's numbers to Jean's.

Between such observers, called inertial observers, Newtonian mechanics preserves a complete relativity, preferring none in any way to any other. Particular problems may dictate particular choice of observer, but only because the choice simplifies the numbers that appear and not because it simplifies the explanations. To describe events taking place on a train you might well decide to travel with the train rather than to stay in the station.

The work begun by Newton and extended by Lagrange and Laplace amongst others dealt with the gravitational properties of matter, mass, and force. The progress was triumphal, especially when applied to celestial mechanics. Indeed by the end of the eighteenth century one of those recurring feelings of disappointment had begun to set in; perhaps all the good problems had been solved and the great theorems already proved. As late as 1842, F. Arago wrote, in his *Eloge de Laplace*:

Five geometers:—Clairaut, Euler, D'Alembert, Lagrange and Laplace shared between them the world of which Newton had revealed the existence. They explored it in all directions, penetrated into regions believed inaccessible, pointed out countless phenomena in those regions which observation had not yet detected, and finally— and herein lies their imperishable glory—they brought within the domain of a single principle, a unique law, all that is most subtle and mysterious in the motions of the celestial bodies. Geometry also had the boldness to dispose of the future; when the centuries unroll themselves they will scrupulously ratify the decisions of science. (Quoted by Struik 1967, p. 137.)[1]

We must remember that geometry was a frequent synonym for mathematics throughout this period, which only makes Arago's pessimism the more striking.

Magnetism and electricity

In the nineteenth century enquiry also began into the twin forces of magnetism and electricity. Faraday described the action of a magnet in terms of lines of

---

[1] These remarks also indicate something of French attitudes to Non-Euclidean geometry at this time.

force that emanate from it, and Maxwell was able to express the laws of motion of bodies in an electromagnetic field. Maxwell's equations imply in particular that electromagnetic radiation is propagated at a constant velocity, which Maxwell identified with the velocity of light in a vacuum, and this velocity, call it c in some units, becomes the maximum speed a signal can be propagated. Information is propagated at this speed at best (sound for instance is vastly slower) and so it is that the news of distant events takes time to reach us. One wonders when God finds out, and the Newtonian answer would be that he knows straight away about every event. Jean and Jo should be able to agree when events took place and about their time order. (Can they?)

However, the theory governing the use of Maxwell's equations does not satisfy the principle of Newtonian relativity outlined above. To obtain the same results predicted for (a) a conductor moving with respect to a stationary magnet and (b) a magnet moving with respect to a stationary conductor recourse must be had to different equations: one kind for situation (a) and one for (b). Yet how can (a) and (b) be told apart; have we here an example of absolute motion? Must we say that in (a) the conductor really is moving and in (b) the magnet is, and if so what kind of knowledge about God is this? Hertz did indeed justify the choice of equation in terms of what was actually moving; as Einstein was to say in 1905, the theory contains asymmetries which did not occur in the phenomena.[2]

Einstein resolved the paradox by reformulating the concept of simultaneity of events, and his train experiment has become famous. Jean stands in the middle of a very fast train, with light sources A and B at the front and back of the train. Jo stands on the track. At the instant Jean passes Jo both receive a flash of light from A and from B. When do they describe A and B as flashing?

Take Jean first. At rest with respect to, and equidistant from, A and B she will say that A and B flashed at the same time. Now take Jo. Light takes a finite time to travel, and so arrives after it sets off. At any instant before the light reaches her A is nearer to her than B and so she must conclude that B flashes first.

Although we may consider Jo moving past a stationary train (she is still nearer A than B until the light reaches her) the precise description of the motion reveals that the concept of simultaneity is relative to the observer. Jean and Jo do not agree as to the 'when' of events, any more than they need to agree as to the numbers describing the 'where', but we can still look for a simple transformation between the descriptions they offer if they only differ by a constant relative motion. Such a transformation was offered by Einstein, and it removes the asymmetry in the theory that he deplored. What emerges is that observers in a state of relative uniform motion do not agree as to the

---

[2] See the start of his paper on special relativity: *On the electrodynamics of moving bodies*. Zur Elektrodynamik bewegter Körper. *Ann. Phys.* **17**, 1905.

simultaneity of events. One sees a sequential order that the other does not, although within limits, as we shall see.

The 'relativity of simultaneity' arises from the finite velocity of light. By the end of the nineteenth century the propagation of light was itself coming under scrutiny. It was supposed to be transmitted through space via a medium called the ether. This useful medium had no tangible properties except to transmit the light; it was the very stuff of absolute space, so to speak, and ideally objects passed through it without affecting it in any way. However, velocity in the ether was for that reason absolute, velocity relative to God, but unhappily experiments seemed to dictate otherwise. George B. Airy had already noticed no difference in the angle of aberration of a telescope viewing a star from a moving earth when the telescope contained air or water (the discussion here follows Holton (1973, p. 264)). If, however, light is propagated through an ether a difference was to be expected. Fresnel therefore proposed that the water drags the ether along in some way; the quantitative consequences of his theory accounted for Airy's results—and also for related results involving refraction obtained by Arago—and were successfully tested in separate experiments by Fizeau. The experiment did imply, however, that air in motion should not affect the ether, so the motion of the earth round the sun should not matter. H. A. Lorentz built a substantial theory on this hypothesis which commanded general acceptance around the turn of the century.

Ether drift

It was Michelson who looked most persistently for ether drift—direct evidence of the relative motion of the earth relative to the ether. If you think of light as a wave and remember that two wave patterns when superimposed yield novel patterns if they are out of phase, then you have in principle a very subtle instrument for measuring small changes: the interferometer. Michelson conceived and built one which was designed to compare the speed of light in two perpendicular directions, a brilliant achievement. However, the results in 1881 were to cause nothing but distress.

If the earth is moving relative to the ether, then the different paths of light from the source down the arms to a mirror and back to the source should affect the light differently. However, to within experimental error, no difference was detected. The results were consistent not with ether drift but with ether drag. Michelson called the result a failure, and the repetition of the same null result with Morley in 1887 caused him to abandon the project. In 1897 he tried again, hoping that the ether drift would vary with altitude, but again nothing showed up. As late as 1927 Michelson and Lorentz amongst others attended a conference on the Michelson–Morley experiment, for it has to be said that the detailed theory of the interferometer in motion is complicated and even Michelson was capable of making mistakes. Until the 1920's the result was one to be 'explained away', as Oliver Lodge had put it in 1893. (These points are taken up and amplified by Holton (1973, p. 266).)

168 *The question of absolute space*

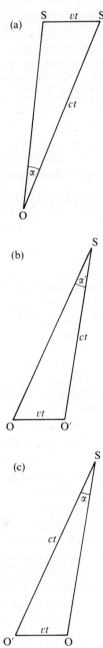

Fig. 17.1. (a) The source S radiates light which reaches the observer at O by which time the source is at S′. The angle of aberration S′ÔS = α represents the difference between the true and the apparent position of the source. (b) However, as Euler noted in 1739, if the measured velocity of light depends on the velocity of the observer OO′, then it is the shorter side O′S which is of length ct and $\alpha' \neq \alpha$. (c) The problem disappears if it can be assumed that light has a constant velocity.

The question of absolute space 169

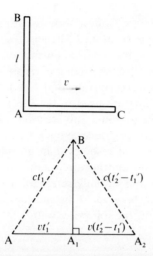

Fig. 17.2. The Michelson–Morley experiment. Light is sent down two tubes AB and AC of equal length l. For simplicity assume that the apparatus is moving in the direction AC with velocity v. For the tube AB we have the following. Light emitted at A at time $t = 0$ hits B at time t_1' when A has moved vt_1' to A_1; the light returns to A in time t_2' and $A_1A_2 = v(t_1' - t_2')$. By Pythagoras, on the journey out

$$(ct_1')^2 - (vt_1')^2 = l^2$$

i.e.

$$(c^2 - v^2)t_1'^2 = l^2$$

and for the return journey

$$(c^2 - v^2)(t_2' - t_1)2'^2 = l^2.$$

So, eliminating t_1',

$$t_2' = 2l(c^2 - v^2)^{-\frac{1}{2}} \simeq \frac{2l}{c}\left(1 + \frac{v^2}{2c^2}\right)$$

For the tube AC light emitted at time 0 hits C at time t_1 and returns to A at time t_2. For the outward journey $l = (c-v)t_1$, and for the return journey $l = (c+v)(t_2 - t_1)$. So, eliminating t_1,

$$t_2 = \frac{2l}{c}\left(1 - \frac{v^2}{c^2}\right)^{-1}.$$

Therefore the theory predicts, to second order in v/c,

$$t_2' - t_2 \simeq \frac{2l}{c}\left(1 + \frac{v^2}{2c^2} - 1 - \frac{v^2}{c^2}\right)$$

$$= \frac{lv^2}{c^3}.$$

If v is the velocity of the earth then v^2/c^2 is about 10^{-8} in metric units.

It would be attractive to suppose that Einstein's theory of special relativity, begun with his paper of 1905 mentioned previously, sprang from an attempt to explain the Michelson–Morley result, and that the atte...pt was immediately successful. However, as Holton shows, both suppositions are wrong. Einstein does not seem to have studied Michelson's work directly. Rather, as the 1905 paper says and subsequent comments of his confirm, the crucial facts influencing him then were the motion of a conductor in a magnetic field, the Fizeau and aberration experiments, and Lorentz's paper of 1895. What concerned him was the apparent need in the theory of the day for absolute motion, when, it seemed to him, only relative motion was conceptually necessary or philosophically desirable. Now Lorentz in 1895 does refer to the Michelson–Morley result, and resolves it in an admittedly *ad hoc* way by supposing objects in motion contract along a line of flight, but the experiment is only one among many that Lorentz refers to. It is much more prominent in the 1904 paper, which, however, Einstein does not appear to have read until after his own was out the following year.

Absolute space

This so-called Lorentz contraction is intended precisely to compensate for the effect of the different paths of the light through the ether, and it is not without physical plausibility. Why should motion not squash things? Fitzgerald independently proposed that moving bodies contract. However, as Lorentz noted, even his own 1904 paper failed to achieve the set task of saving appearances by predicting them correctly, despite a battery of saving auxiliary hypotheses. It saved the Michelson–Morley experiment but failed to make Maxwell's equations invariant even for small speeds. Furthermore, Lorentz's paper predicts different results from those of Einstein concerning an optical Doppler effect and stellar aberration. The two theories are not in fact observationally equivalent.

Yet the inherited belief in absolute space continued to hold sway over physicists for a considerable time. Poincaré, who had turned to physics after a brilliant career in mathematics, never accepted Einstein's idea of relativity and died in 1912 believing that the classical theory could be repaired. His paper of 1905 is sometimes considered, with Lorentz's, as the best attempt to do just that. Poincaré's position represents an interesting compromise between the contemporary and the new physics. He sought to formulate physical laws in a relativistic way, inasmuch as all references to absolute space and absolute time were to be abandoned and only relative motions considered, and explicitly sought forms for the laws which would be invariant under transformations from one observer to another. None the less he sought to explain the apparent validity of absolute laws and to maintain the privileged position of the ether.[3] Nor did Lorentz ever quite reconcile himself to the new

---

[3] Poincaré: 'Sur la dynamique de l'électron' *Rend Circ. Mat. Palermo*, **21**, 129–76 ((1905), published 1906), *Oeuvres*, vol. 9, pp. 494–551. Holton (1973, p. 187) also establishes the influence of Poincaré's charming book *La Science et L'Hypothèse* (1902) on Einstein.

theory, and Michelson never did, remarking once to Einstein that he was sorry his own work may have helped to start this 'monster'.[4] With these influential names weighing in against the new theory its progress was understandably slow.

What stuck for most people was the thorough abandonment of absolute space. The existence of the external world independent of the observer had become associated with a belief in absolute properties of the objects in space. They *do move, they are* there, whether we like it or not. Einstein showed, however, that the description we offer is dependent on what we are doing and where we are in a more subtle way than any one had expected, and more subtle than many were prepared to accept. What made for final acceptance of the theory was, at least initially, a philosophical predilection. German writers on natural science were always more philosophical than their English or French counterparts; they paid great attention to their presuppositions and were inclined to emphasize how their science flowed from general, abstract considerations (compare Duhem's (1954) distinction between 'ample' and 'deep' minds). The nature of space and time, of scientific method, and epistemological and ontological considerations animated influential writers like Helmholtz and Mach and surely prepared the way for Einstein's revolutionary treatment. By contrast, the commonsense empiricism of English and American workers inclined them to reject Einstein's work because it gave them different aesthetic standards. If you accepted Einstein's theory you did so because of its elegance and beauty since the experimental evidence was so contentious, and if you rejected it, it was because it was far-fetched and disturbing. One aspect of a theory's aesthetic is its mathematics. Here a decisive service was paid by Minkowski's lecture in 1905, subsequently published many times (e.g. in *The principle of relativity* 1923 (reprinted 1952)). Here, as he triumphantly announced,

three dimensional geometry becomes a chapter in four-dimensional physics... Space and Time are to fade away into the shadows, and only a world in itself will subsist (p. 80).

It was a world of space–time, for as Minkowski also said:

No-one has yet observed a place except at a time, nor yet a time except at a place (p. 76).

Put like that it is unarguable, but it dethrones God from his sensorium of space in which time passes.

Minkowski's four-dimensional world is both simple enough to be easy to use, with practice, and novel enough to free us from the shackles of our

---

[4] R. S. Shankland, *Conversations with Albert Einstein*, p. 52, quoted by Holton (1973, p. 317).

172 The question of absolute space

absolutist inheritance. We shall turn to it in the next chapter. There is first one more important experiment to describe.

The Kennedy–Thorndike experiment

Even if the Michelson–Morley experiment is given its modern interpretation it only affirms that the spread of light is isotropic, i.e. the same in all directions for any inertial observer. It does not assert the stronger result that the speed of light has the same numerical value for every such observer. The experiment we now take to confirm that result is the Kennedy–Thorndike experiment of 1932. An interferometer with two arms of unequal length was used, and was set up with great care being taken to ensure that it did not change its size significantly; it was mounted on quartz, a particularly stable solid, in a vacuum kept to a temperature that was constant to ± 0.001 °C. The two intertial frames are the apparatus at one point in the earth's orbit and the same apparatus six months later when the earth has reversed its velocity with respect to the fixed stars. At any stage the interferometer, which now operates with continuous monochromatic light, records a pattern which is a measure of how much longer it takes the light to travel the extra distance by which the longer arm exceeds the shorter arm (strictly, that distance is traversed twice, there and back, so the extra time needed is $2(l_1 - l_2)/c$). The apparatus was run over long periods, up to a month, and the effect of reversing the earth's velocity was estimated since an uninterrupted six months run was not possible. The result was null, to within experimental error. (An excellent account of the experiment, with historical notes, can be found in Taylor and Wheeler (1963, p. 78) on which this account is based.) However, the relativistic interpretation of the experiment is not forced upon us. As Taylor and Wheeler (1963, p. 80), observe, in 1932 it was possible to believe that (i) a Lorentz–Fitzgerald contraction of lengths takes place with respect to absolute space, or (ii) a Lorentz–Fitzgerald contraction takes place absolutely and clocks are also slowed absolutely, or (iii) special relativity holds.

Theory (i) predicts an observable difference in the interference patterns large enough for us to reject on the basis of the results obtained, but theory (ii), like special relativity, predicts no observable effect. The clocks seem to be slowed just the right amount by the motion. Therefore absolute changes in lengths and times are allowed by both Michelson–Morley and Kennedy–Thorndike experiments. Relativity theory prefers the equivalence of all inertial observers, which these experiments also confirm.

What Lorentz had had to postulate was the absolute contraction of a body along the line of flight. A second modification to the theory would be an absolute dilation of time. Einstein preferred to postulate two hypotheses: the equivalence of all physical systems as described by two observers subject to no forces (inertial observers) and the constancy of the velocity of light to all such observers. These two postulates constitute the principle of relativity as it is applied, the equivalence being the pure 'principle of relativity'. They differ from the *ad hoc* modifications in supposing that the changes are relative and not absolute in nature.

On scientific enquiry

We are now almost ready to follow Einstein's first description of space and time in its geometrical aspects. The theory is now known as special relativity, special because it lacks any reference to gravity and the action of forces. The more general theory is deferred till later (p. 196).

However, I should like briefly to turn to the fascinating question of the nature of scientific enquiry itself. There are a number of theories of scientific method about, all of which conflict to varying degrees, for instance those of Kuhn, Lakatos, Putnam, Feyerabend, and Holton (see Bibliography). What emerges from these critiques, and which should have been evident to all working scientists, is that a scientist's allegiance is first of all to his theory and not to his facts. Once an experiment conflicts with the theory the impulse of every scientist is to check the experiment and to check his calculations. Even if a thorough check does not explain away the anomaly, the scientist does not consider the theory to be refuted. Often the experiment is considered to be a failure, witness Michelson in 1887. Lorentz wrote to Lord Rayleigh (18 August 1892)

Fresnel's hypothesis . . . would serve admirably to account for all the observed phenomena were it not for the interesting experiment of Mr. Michelson, which has, as you know, been repeated after I published my remarks on its original form, and which seems decidedly to contradict Fresnel's views. I am totally at a loss to clear away this contradiction, and yet I believe if we were to abandon Fresnel's theory we should have no adequate theory at all . . . Can there be some point in the theory of Mr. Michelson's experiment which has as yet been overlooked. (Quoted in full by Schaffner (1972) and here abbreviated.)

We must here distinguish two kinds of experimental work. One kind, by far the most common, takes a theory as given and calculates something on the basis of that theory: a mass, a charge, a lifetime, a phase-shift, or whatever it is. There is here no attempt even in principle to refute the theory; an anomalous result is always put down to the apparatus, which can be very complex and liable to faults, as, for example, an interferometer is. The second kind is an attempt to prove a theory but in the event of a provocative result the theory need not fall. It can be propped up by extra hypotheses; a particular example is given above (p. 163). Furthermore, this second kind of experiment (the 'crucial' experiment) is only performed when there are competing theories with rival predictions. The experiment is then taken to pick the winner or to reveal the hidden extra hypotheses the loser needs to stay alive, according to taste. However, it is nearly always the theories that come out as what matters to scientists, not the experimental results alone, and this is for the reason that an experiment cannot be performed, or even conceived, outside of a theory. It is his theory which structures the world of the experimenter, and which governs his interpretations of what he can do and has done. An experiment like Michelson–Morley can be interpreted, for instance, as strange behaviour of the ether, or as reasonable behaviour of light according to Einstein's theory of special relativity, but it cannot be simply 'read' as the 'facts'.

It is also worth remarking that to design an experiment is to commit yourself to a lot of theories apart from the one you want to check, typically for instance optics, electronics, and something newish, like ultra-low-temperature physics, to which your equipment is supposed to conform. You then 'see' the experimental facts that your theories permit you to conclude from your measurements, and so potentially a chase through many theories begins, and with it the possibility of '*ad hocness*'.

Therefore it is not surprising that theories are what survive; theories are what everyone works with. A decisive experiment would need to summon up a new theory for it to be truly intelligible, and not intelligible but inexplicable on the old one. An anomalous result on this formulation is a sign that we do not fully understand our theory and concepts, and that perhaps the concepts and theory need changing. Seen this way it is not surprising that 'crucial' experiments are in fact few and far between; they are always weighed in the balance against a whole system of thought and need a new system to tip that balance their way—witness the philosophical appeal of relativity (to some). Indeed, it seems that a scientist proceeds by structuring all his experience according to a set of theories, some of which may be of a very general character and some of which can be very specific. At their vaguest they are simply an acceptance of certain sights as what they seem to be. This level is unconscious and common to most of us and can scarcely be called a theory at all. At a higher level is, for instance, the learned perception of depth in a photograph, of perception as 'what it really is'. Then there are generally agreed criteria as to the persistence of objects: the train entering and then leaving a tunnel is the same train. This is perhaps the first level of abstraction which could be made explicit as a philosophical theory, and our scientist has scarcely begun yet to operate as a scientist. It is because his theories are grounded in ordinary language, however, that 'pure facts' were ever thought of. He begins to be a scientist when he structures a specific part of his experience according to his scientific training, where both the subject matter and the methods of study are special.

In the case of relativity theory we should recognize that it was pursued and adopted not simply because it explained the known 'facts' or because it predicted new ones. For the most part its novel consequences can be explained away one by one with (different) modifications of the old theory, although a Lorentz-type theory to account for all of them at once would be a ragged thing indeed. Its general acceptance did not come until at least the 1930's, which would seem to undermine any such simple theory of scientific progress. We must also reckon with such subjective factors as elegance and naturalness. Just as non-Euclidean geometry requires for its full development a plausible interpretation as the geometry on a certain surface, so Einstein's theory needed a commitment to the operational redefinition of the basic geometrical terms length and time. Only when the physical community had had time to readjust did relativity begin to seem 'right'.

18 Space, time, and space–time

The description of space–time.

Imagine, if you will, a satellite in circular orbit around its parent planet. At regular intervals the two are photographed, obtaining a recording looking something like Fig. 18.1(a). If we superimpose the photographs we see all the positions at once, obtaining a record of where the satellite has been (Fig. 18.1(b)). However, now we have lost record of when it was there and how fast it was going. Therefore we look for a way of recording the information which preserves both the position of the satellite and when it was there. Such a way is evidently to stack the photographs evenly in time order and look at them all. We shall arrange them along an axis calibrated 1 h, 2 h, etc. and so, centring them on the fixed planet, we obtain a record like Fig. 18.1(c). We can reasonably infer the position of the satellite between times, and when we re-draw the information without the frames of the photograph we obtain a trace like Fig. 18.1(d). Implicit in the photograph was a scale of distance, which we may consider in terms of axes, an x axis across the photograph and a y axis up it. This means that the planet itself, for instance, is always at (0, 0). A time axis is provided by the arrangement of the photographs, so we have three axes for the complete record of the orbit: x, y, and t along the page.

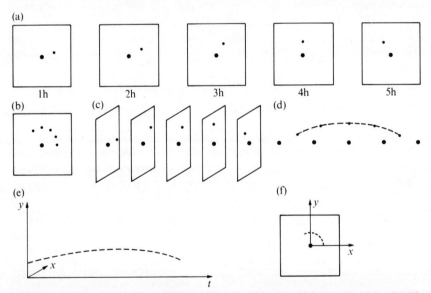

Fig. 18.1. (a) The hourly record of a satellite in orbit around its parent body. (b) The photographs superimposed. (c) The photographs arranged in order. (d) The full record of the orbit. (e) The graph of the orbit. (f) End-on views of the graph, suppressing t.

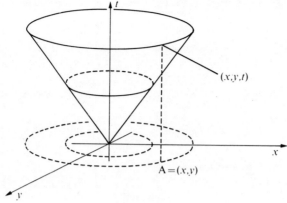

Fig. 18.2.

Such a trace is called the space–time graph of the satellite. In this example it is three dimensional, since the orbit itself lies entirely in the (x, y) plane. In general a body moves in three space dimensions, and so needs four dimensions for its space–time graph: x, y, and t. Such pictures are hard to draw, and we shall frequently suppress a space axis in order to obtain simpler pictures. An effort of the imagination can always supply the missing dimension. There is, however, one feature of these graphs which might be uncomfortable. The planet has always been at rest, but because it appears in each photograph it appears as a line on the graph—in this case lying along the t axis. This feature is inescapable: stationary objects leave a path on the graph, which, however, has unvarying x and y (and z), and only t varies. If one drew the planet for a fixed value of x, y, and t it would appear as a point existing at that moment of time only and at no other. A line parallel to the t axis represents a motionless object enduring through time. A single point with unique x, y, (z), and t co-ordinates is called an event to highlight its momentary character.

Naturally, these graphs have much in common with the usual ones showing position as a function of time. If you consider the (y, t) plane and the horizontal projection of the path upon it, the slope of that path is a measure of the velocity of the body whose motion it represents. However, they have other properties which are less familiar.

Let us consider a flash of light at the origin and see how the light spreads out over the universe. We know, of course, that the wavefront spreads out at constant speed from the source, like a ripple in a pond, and any part of it moves outwards in a straight line. Representing space in two dimensions, x and y thought of as lying in a horizontal plane, and putting t vertically upwards, we would draw the space picture as successive concentric circles. The space–time picture is obtained by stacking the space pictures at the appropriate time level. The ripples in the space picture are translated into

178 Space, time, and space–time

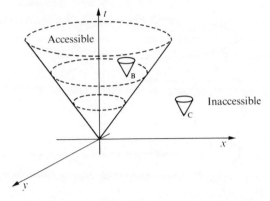

Fig. 18.3.

sections of a cone in the space–time picture. The space–time picture of the advancing light is that of a cone, which is called the light cone and has a great physical significance.

Imagine that the light is intended to reach someone in space at A, say, who has a definite space address (x, y). It reaches him at time t, and this is symbolized in the event (x, y, t) in the space–time picture, where his path (vertical and parallel to the time axis because he is motionless) first meets the light cone. Accordingly we can divide up space–time into two regions, separated by the cone itself, corresponding to the time it takes the light to reach A. The region outside the cone, including that part of A's path below the cone, corresponds to those parts of space–time which have not received the light. The region inside the cone contains those points which have received the light. Now, if we think of that light as information, this means that regions outside the cone cannot be reached by information from O. This is not to say that regions of *space* cannot see the light; of course they can, but only at a certain *time* and not before (or after, if it is indeed a flash). All regions of space eventually perceive the flash, but at different times as the wavefront spreads out. Suppose that on seeing the flash A will set off his own. If he does so too early, before the light reaches him, we would not say that the flash at O caused the flash at A. If, however, he first sees the flash and then triggers his own, then we could say that the flash O caused the flash at A. In space–time terms A flashing is an event with co-ordinates (x, y, t). If that event lies outside the cone with vertex at O it was not caused by the flash at O, but if (x, y, t) is inside the cone it could have been. The region outside the light cone therefore is made up of events which could not have been caused by anything at O.

The region inside the light cone is, by contrast, made up of events which can see O and so be triggered off by light, or some slower signal, reaching them from O. Therefore the light cone divides space–time up into events which are causally independent of anything happening at O (outside the cone) and those which are possibly causally dependent on it. Again, if we agree that the effect must always precede the cause, we classify events into the undetermined in time (outside the cone) and the determined in time order (events inside the cone, which represent the future of the signal).

Just as an event at O can cause subsequent events elsewhere, so the event O can be regarded as a possible response to certain previous events. Again, not all events can be considered, but only those visible at O. Thus, if the sun suddenly switched off it would be a few minutes before we could know and do anything about it, because that signal would take that time to reach us. Indeed, the sun is so large that the last we would see of it would be a black spot which would swallow it entirely in two and a half seconds.

Those events which contribute to what happens at O, and which we see for instance when we look into space, lie symmetrically with the light cone that we have drawn before and form the interior of the backward light cone. The cone itself is all the light falling on O at that moment, and the further the light has travelled the further back in time lies the information it carries. Outside the cone are all the events invisible from O and so not capable of affecting O. (Remember that O, being a point on the diagram, is itself a momentary event and not a place.)

Each point in space–time, each event, carries with it its own forward and backward cones, its own future and past.

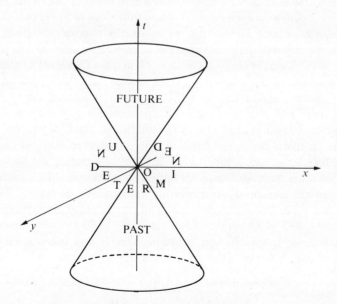

Fig. 18.4. The forward and backward light cones at O divide space–time into three regions: future, past, and undetermined.

Clocks and surveying

We consider next how such a description of space–time might be established in reality, that is how a survey of a region of space–time might be made. We are in the position of the beings on a surface. Any survey we make of space–time must be intrinsic, since we cannot step outside of the space–time we inhabit. Let us employ a convention: our meter rod carried about remains everywhere the same length (we examine this convention later). In this way, we cover space with a three-dimensional lattice calibrated in metres. Each point of this lattice is equipped with a clock to record the time co-ordinates of an event, and we must therefore consider what a clock is and how they can be synchronized on the grid. We follow the description given by Taylor and Wheeler (1963).[1]

Light travels with constant velocity, whatever the velocity of source and observer.[2] This experimental result allows us to build very simple clocks which can be fitted naturally into an exposition of the theory and into space–time diagrams. Therefore first let us consider the clock design that we shall call the standard clock. A light source at A gives out a flash of light, which travels down a tube AB 0·5 m long. It is reflected at B and returns to A, where it is counted and re-reflected. The total distance travelled is therefore 1 m and the elapsed time we shall call one unit of time, or 1 m of light time. In this way the measurement of time is reduced to measurement of length, and the same unit, the metre, is used to measure both. We have again appealed to our convention regarding metre rods.

The next step is to place a clock at each grid point in the lattice to record the time component at each event in space–time. This requires that the clocks are all synchronized. To do this we arbitrarily choose one clock, C_o, to represent the origin of the grid, and set all neighbouring clocks in the following fashion: with all clocks stopped the clocks 1 m from C_o are set to read 1 m of light-time ahead of the time on C_o, those clocks 2 m from C_o are set to read 2 m of light-time ahead, and so on. C_o is then started and a flash is set off through space which triggers each clock into motion as it reaches it. In this way a synchronized grid of clocks is established which represents our co-ordinates on space–time. Any event in space–time is recorded on the nearest clock to fix its position, a superlative standard of accuracy for astronomical work. For other work smaller grids with finer clocks would be used.

One pleasant feature of our measuring units calls for comment. The speed of light is 1 m per metre of light-time in all frames. Therefore it appears in a space–time graph as a straight line bisecting the angle between the axes. It is straight because it travels with constant velocity and is an angle bisector because of our choice of units, not for any mysterious reason.

[1] A good case can be made, however, that time is both logically prior to space and more readily measurable with 'atomic' clocks.

[2] The experimental result of Kennedy and Thorndike well illustrates how theory laden such statements are. Strictly, only the observer was in motion during the experiment, but a natural relativism has crept into the description of it.

The invariance of distance—the pure space case

Here, of course, we are used to the convention of unvarying rods, so much so that it seems inescapable. On this convention a number of superficial disagreements can still arise between observers which can easily be sorted out. Jo and Jean can give to every object in the survey a pair of numbers which are its x and y co-ordinates. When they come to compare their results, the co-ordinates of each object in the two surveys may differ. It is the matter of a moment to agree on a common origin for the two surveys, but even then the co-ordinates of an object will still differ. The next agreement must be on a common unit of measurement, so that objects which Jean declares to be one unit from the origin Jo also says are so. There is now a surprising connection between lengths and co-ordinates. If both agree that OA is one unit long, they may still disagree as to the co-ordinates of A. O is (0, 0) in each case. If Jean writes co-ordinates as $(\ ,\ )_e$ in her survey, and Jo as $(\ ,\ )_o$ in hers, it might be that A is $(4/5, 3/5)_e$ to Jean and $(5/13, 12/13)_o$ to Jo. In each case $(4/5)^2 + (3/5)^2 = 1 = (5/13)^2 + (12/13)^2$ by Pythagoras Theorem. In general, if a length OB is agreed to be l unit, and agreement follows from the choice of unit, then we have $(x, y)_o$ and $(x', y')_e$ for B and $x^2 + y^2 = l^2 = x'^2 + y'^2$. The square of the interval between any two points is agreed by both of them not only to be l^2, the square of the distance between the points, but to be the sum of the squares of the co-ordinates of the point in each case. Pythagoras Theorem is here interpreted as a physical principle, called the invariance of distance, once agreement has been reached as to the choice of unit.

The disagreement over the co-ordinates in the two systems is readily explicable. If A is $(4/5, 3/5)_e$, the axes for Jean must be as shown in Fig. 18.5,

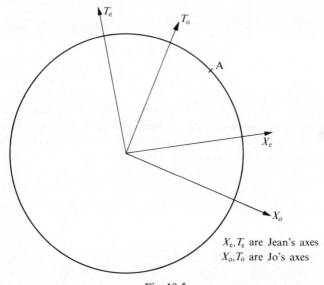

X_e, T_e are Jean's axes
X_o, T_o are Jo's axes

Fig. 18.5.

182 *Space, time, and space–time*

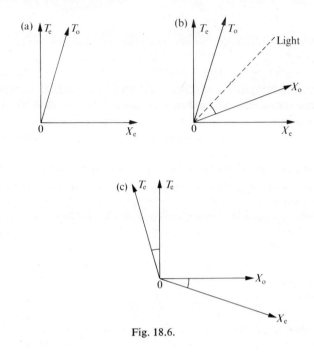

Fig. 18.6.

and those for Jo must be as in the Fig. 18.5 also. We see that the co-ordinates can be made to agree, $x_e = x_o$, $y_e = y_o$ by rotating the axes. Such a rotation does not, however, alter the positions agreed by each to be 1 unit from the common origin O, for they form a circle, centre O. We need to remember this way of reaching agreement when we turn, as we do now, to the more difficult question of surveying space and time.

Other axes

We saw that two surveyors of space might differ in their choice of (rectilineal) axes by choosing different origins, different scales of measurement, and different orientations for their axes. In surveying space–time the choice of origin is entirely arbitrary and agreement can soon be reached. However, a difference between Jean's and Jo's choice of axes for space–time is slightly harder to analyse than it is in the case of pure space. Fortunately it is enough to consider the simplified case of one space axis and one time axis.

Suppose Jean draws her axes in the usual way with X_e at right angles to T_e as shown (Fig. 18.6(a)). Then the time axis T_o for Jo, if it does not lie along Jean's axis must lie skew to it. Now Jo's time axis is represented to her (Jo) by her clock, which is motionless at the origin and enduring in time. By making her origin coincide with Jean's she can arrange for her time axis to have the time = zero event coincident with Jean's, but her estimates of time for other events disagree because of the relativity of simultaneity: what she sees as motionless enduring are the events which lie on her time axis and for which

x_o is constant but t_o varies, and Jean registers these as a sequence of events for which x_e and t_e both vary. The events on Jo's time axis are, to Jean, events of a moving object—in fact, Jo. In this way they record the difference in axes as corresponding to the physical difference of uniform relative motion. However, they both agree that the velocity of light is 1 in each frame, so x_o appears on Jean's axes as shown in Fig. 18.6(b). The two reference frames appear not to be rotated with respect to each other but squeezed by an angle which measures their relative velocity. The situation is perfectly symmetrical; had we begun with Jo and looked upon Jean and her axes as new we would have seen her move, but in the opposite direction as Fig. 18.6(c).

There remains the question of the choice of scales that the two can make. In the pure space case a difference in scale led to different sets of points forming circles one unit from the origin; points Jean considered to lie on a circle of unit radius were not so considered by Jo until both agreed on a common standard of length. In the case of the space–time universe, what is it that provides such a check? Let Jo travel past Jean with velocity v, carrying with her what she (Jo) considers to be a standard clock of length 0.5 m— strictly 0·5 m_o since this estimate of length is Jo's. Jean sees the length contracted to a length λ say. We read this into the space–time graph as follows: successive positions of Jo's clock are recorded and thus her passage through space–time, as measured by Jean. The path of light in Jo's clock is also marked. We are interested in Jean's assessment of the event, marked as B, which Jo would call one beat of her clock and in the co-ordinates of B in the two frames.

Jean reads A as $(\lambda, 0)_e$ because of the contraction due to movement. When the light has travelled down the clock Jean records the event as $(\lambda/(1-v), \lambda/(1-v))$, and she records B as $(v\lambda/(1-v^2), \lambda/(1-v^2)_e)$. (The calculation is done in the caption to Fig. 18.7.)

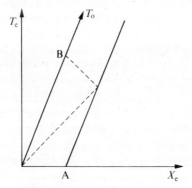

Fig. 18.7. To Jean the path of O is $x = vt$ and the path of A is $x = vt+\lambda$. The path of light from O outwards is $x = t$ and the path back is $x+t = b$, where b is a constant depending on the co-ordinates of the event which is the reflection of the light in Jo's clock: $(\lambda/(1-v), \lambda/(1-v))_e$ to Jean. So the return path is $x+t = 2\lambda/(1-v)$ and the point of return B is $(2v\lambda/(1-v^2), 2\lambda/(1-v^2))$.

Jo, of course, considers the event B to mark the first return of the light, or one beat of her clock. So for her the co-ordinates of B are $(0, 1)_o$. Her X_o co-ordinate is zero because the light has returned to where it began, the left-hand end of her clock, which is stationary for her and marked the agreed origin of co-ordinates for her and Jean when $t_e = 0 = t_o$.

Neither space nor time co-ordinates agree separately in the two systems. In particular, Jean records the time co-ordinate of B as $2\lambda/(1-v^2)$, which is larger than 2λ, indicating that she records Jo's clock as beating slow. This is the phenomena of time dilation in which moving objects are recorded as slowed down in time. Jo's clock beats 'one' later than Jean might have expected it to. What is preserved is a twisted analogue of the distance in pure space, for in each case $t^2 - x^2 = 1$. $t_o = 1$ and $x_o = 0$ evidently; $t_e = 2\lambda/(1-v^2)$ and $x_e = 2\lambda v/(1-v^2)$ so

$$t_e^2 - x_e^2 = \frac{4\lambda^2(1-v^2)}{(1-v^2)^2} = \frac{4\lambda^2}{(1-v^2)}.$$

Remembering† that if a rod of length $\frac{1}{2}$ in Jo's frame is measured by Jean it is found to be contracted and does not appear as $\frac{1}{2}$ m but as $\frac{1}{2}(1-v^2)^{1/2} = \lambda$, we find

$$4\lambda^2 = 1 - v^2$$

and

$$t_e^2 - x_e^2 = 1.$$

Remarkably, both observers agree that the points $(x_e, t_e)_e$ which satisfy $t_e^2 - x_e^2 = 1$, have co-ordinates $(x_o, t_o)_o$ which satisfy $t_o^2 - x_o^2 = 1$. Geometrically these points lie on a locus, a rectangular hyperbola, which is physically meaningful. The quantity $(t^2 - x^2)^{1/2}$ which separates an event from the origin is called the *interval*, by analogy with distance in the pure space case. In general any two events E and F in space–time are separated by an interval, and that interval is independent of the frame used to calculate it or to record the position of the points. Thus if E has co-ordinates (x, t) and F has co-ordinates (x', t') then $((t-t')^2 - (x-x')^2)^{1/2}$ is a constant, whatever observer and frame x, x', t, and t' are measured in. The simple expression we first obtained for the interval is obtained by letting the origin of the co-ordinates be at F, so that $x' = 0 = t'$.

The asymptotes of this rectangular hyperbola represent the limiting position of B for a succession of faster and faster observers, as seen by Jean. No matter how fast Jo goes, B will lie on the hyperbola at a position that is nearer the asymptote the faster she goes. This is as it should be, for the asymptotes are the events on the light cone, the path of a pulse of light emitted from O. The hyperbola and its asymptotes in space–time are the geometrical analogue of the circle in the pure space case.

† Equal times in the Michelson–Morley experiment imply that $t_2 = \frac{2AC}{c}\left(1 - \frac{v^2}{c^2}\right)^{-1} = t_2' = \frac{2AB}{c}\left(1 - \frac{v^2}{c^2}\right)^{-1/2}$, whence $AC = AB\left(1 - \frac{v^2}{c^2}\right)^{1/2}$. See also Exercise 18.2.

Summary

We can compare the space–time surveyors with the space observers in the way shown in Table 18.1.

Table 18.1.

	Space	Space–time
Observers use	Rods	Rods and clocks
Observers might differ in and in unit of	Orientation length	Velocity length (and time)—in each case 1 m once the clocks are standardized
despite differences in	orientation	velocity
observers can agree on	the distance between points $(x^2+y^2)^{1/2}$	the interval between events $(t^2-x^2)^{1/2}$
provided they agree on the unit of	length	length (and time)
They will then agree on	the unit circle of points around the origin	the rectangular hyperbola asymptotes are the light-cone at the origin

In addition the space–time observers will find that to each other they seem contracted in length alone the line of flight and dilated in time. We then have the following cycle of equivalences:

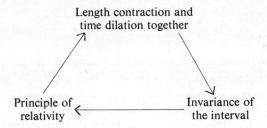

Paths

Suppose that we have to measure the length of a path in space. We can imagine it lying in the (x, y) plane and we might well agree that its length was adequately measured by approximating the curve with straight segments and measuring them in the way we would probably measure the curve with a ruler. Since this method of measurement reduces to measuring lots of segments and adding, it is independent of the choice of axes x, y. This is as it should be, for the length of a curve is surely a property of the curve itself and not of the co-ordinates with which we might happen to describe it.

We can similarly use the invariance of the interval to measure paths in space–time. Notice a curious thing, however: not every path in space–time can represent the path of a physical object. There can only be paths in a space–time diagram for objects travelling no faster than light, for only these

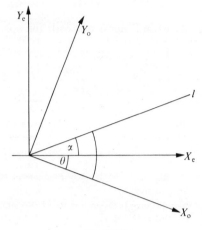

Fig. 18.8.

paths have their slopes at any point within the forward light cone at that point.[3] To measure any such path we again approximate it with linear segments and appeal to the invariance of the interval $(t^2-x^2)^{1/2}$ along each segment, instead of the invariance of distance. The quantity measured along a path in this way is called the *proper time* along the curve. It is the same for all observers because of the way in which it was calculated, and it has the units of time because every part of the curve is approximated by an interval along which t^2-x^2 is positive and so t is greater than x which makes the intervals time like.

Just as we may draw many curves in space between two points, so we may imagine many paths in space–time between two events. Whereas in space the straight line is distinguished as the shortest curve between two points, in space–time the reverse is true and straight lines measure the longest interval between two events. The reason is time dilation. Along sloping lines (representing motion) intervals appear longer than they do against vertical ones (representing rest relative to the observer) since it is t^2 *minus* x^2 we are interested in. So the total interval along a curve is less than along a vertical line joining the same two events. The same happens when a constant velocity and an accelerating–decelerating man compare their interval measurements. Proper time is maximized along straight paths in space–time.

However, let us redeem this excursion into co-ordinate geometry. There are several supposed paradoxes in special relativity and I shall discuss them in the next chapter, where we shall find that the space–time diagrams help us to avoid many mistakes.

---

[3] The slope of a path measures the velocity of the object on the path.

Appendix

The number-crunching aspect of special relativity can be made easier by the introduction of the same hyperbolic functions that we met in connection with non-Euclidean geometry. We have already seen that the invariance of the interval is a fundamental property common to all observers, or, more precisely, that it is agreed upon by all observers travelling with constant velocity.

To look at the pure space case first, Jean and Jo agreed on $x_o^2 + y_o^2 = x_e^2 + y_e^2$ although Jo's and Jean's axes differed by a rotation through θ, say. The relation between x_o, t_o, x_e, and t_e is

$$x_o = x_e \cos\theta - y_e \sin\theta$$
$$t_o = x_e \sin\theta + y_e \cos\theta$$

which preserves the distance $x^2 + y^2$ since

$$\begin{aligned}x_o^2 + y_o^2 &= (x_e \cos\theta - y_e \sin\theta)^2 + (x_e \sin\theta + y_e \cos\theta)^2 \\ &= x_e^2(\cos^2\theta + \sin^2\theta) + y_e^2(\sin^2\theta + \cos^2\theta) \\ &= x_e^2 + y_e^2\end{aligned}$$

A line l inclined at an angle α to Jean's x axis is inclined at an angle $\theta + \alpha$ to Jo's x axis, and so we say that the angles transform additively. Of course they do not agree about the slope of the line l. To Jean the slope is $\tan\alpha = y_e/x_e$ and to Jo, $y_o/x_o = \tan(\theta + \alpha)$. The connection between them is

$$\tan(\theta + \alpha) = \frac{\tan\theta + \tan\alpha}{1 - \tan\theta \tan\alpha}$$

which is not additive; we do not have

$$\tan(\theta + \alpha) = \tan\theta + \tan\alpha.$$

How does this help us with the space–time computations?

Let us compare two observers, both momentarily at O, Jean at rest and Jo moving. An event E is recorded by Jean at $(x_e, t_e)_e$ and by Jo at $(x_o, t_o)_o$. They have already agreed a common origin O and a value for the interval between O and E: $t_e^2 - x_e^2 = t_o^2 - x_o^2$, although $x_e \neq x_o$, $t_e \neq t_o$. The relation between the co-ordinates turns out to be

$$x_o = x_e \cosh\theta + t_e \sinh\theta$$
$$t_o = x_e \sinh\theta + t_e \cosh\theta$$

and this preserves the interval, for

$$\begin{aligned}t_o^2 - x_o^2 &= (x_e \sinh\theta + t_e \cosh\theta)^2 - (x_e \cosh\theta + t_e \sinh\theta)^2 \\ &= t_e^2(\cosh^2\theta - \sinh^2\theta) - x_e^2(\cosh^2\theta - \sinh^2\theta) \\ &= t_e^2 - x_e^2 \quad \text{since} \quad \cosh^2\theta - \sinh^2\theta = 1.\end{aligned}$$

θ has a physical interpretation. Jean regards T_o as Jo's flight path, and Jo's velocity is $\tan\theta$, where we must have $\tan\theta < 1$, the velocity of light in a

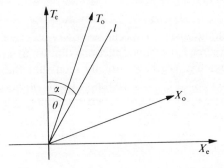

Fig. 18.9.

vacuum, and so $\theta < \pi/4$. However, $\tan \theta$ is an awkward measure of velocity to use because, as we have just seen, it is not additive, and we cannot use the angle itself because of the restriction $\theta < \pi/4$. Let us take an example. If Jo approaches Jean at three-quarters the speed of light and fires a rocket ahead of her which she registers as travelling at three-quarters the speed of light, the speed Jean sees the rockets approach her at is not $\frac{3}{4} + \frac{3}{4} = 1\frac{1}{2}$ times the speed of light but only 0·96 of the speed of light. Indeed it turns out that if v_o is the velocity of Jo relative to Jean and v_r is the velocity of the rocket relative to Jo then $(v_o + v_r)/(1 + v_o v_r)$ is the velocity of the rocket relative to Jean.

In our example $v_o = \frac{3}{4} = v_r$ and

$$\frac{v_o + v_r}{1 + v_o v_r} = \frac{\frac{3}{4} + \frac{3}{4}}{1 + \frac{3}{4} \times \frac{3}{4}} = \frac{24}{25} = 0 \cdot 96.$$

Therefore we look for a velocity parameter which depends on the velocity but is additive. Such a parameter, by analogy with $v = \tan \theta = \sin \theta / \cos \theta$ should be ϕ, where $v = \tanh \phi = \sinh \phi / \cosh \phi$. Indeed we find that if $v_o = \tanh \phi_o$ and $v_r = \tanh \phi_r$, then

$$\frac{v_o + v_r}{1 + v_o v_r} = \frac{\tanh \phi_o + \tanh \phi_r}{1 + \tanh \phi_o \tanh \phi_r} = \tanh(\phi_o + \phi_r).$$

Therefore the *velocity parameter* ϕ defined by $v = \tanh \phi$, is additive.

Caution: ϕ is *not* the angle between the path of the rocket and the T axis. It is not visible in the picture at all.

However, there is one gain. The function tanh has agreeable properties: since $\cosh \phi$ is always greater than $\sinh \phi$, $\tanh \phi = \sinh \phi / \cosh \phi$ is always less than unity. Also, $\cosh^2 \phi - \sinh^2 \phi = 1$, and so as ϕ approaches $\pi/2$ and $\cosh \phi$ and $\sinh \phi$ increase, we have $\tanh \phi \to 1$.

Therefore, as the velocity parameter increases without limit ($\phi \to \infty$), the velocity approaches unity ($v \to 1$). It seems appropriate that the velocity parameter should not only be additive but unbounded.

Furthermore, if ϕ is small, say below 1/1000, tanh ϕ is very nearly ϕ. That is, $v = \tanh \phi \simeq \phi$ and the velocity parameter nearly agrees numerically with the velocity. It follows that at slow speeds and to a certain level of accuracy—again one part in a thousand—velocity is additive. The speeds we have in mind are below 186 mile s^{-1}, which is about 750 000 mile h^{-1}. However, this is fast enough to show that velocity is additive to a high level of accuracy at ordinary velocities.

Sinh, cosh, and tanh: the analogy with hyperbolic geometry is beginning to come out and will be explored further as we proceed.

Exercises (mathematical)

18.1 Earlier, physical reasons were given for only considering linear transformations between Jean and Jo. Attempt to give a mathematical derivation by showing that the only analytic transformation $f(x, y) = (x_1, y_1)$ such that $x^2 - y^2 = x_1^2 - y_1^2$ is $x_1 = ax + by$, $y_1 = cx + dy$ for some constants a, b, c, and d. What is wrong with the transformation:

$$x_1 = \frac{x^2 + y^2}{(x^2 - y^2)^{1/2}}, \quad y_1 = \frac{2xy}{(x^2 - y^2)^{1/2}}?$$

18.2 Jean detects an arbitrary event E by sending a pulse of light to it along $t_e - x_e = T_e$ and receiving the pulse back along $t_e + x_e = T_e'$. Since E is a momentary event, Jo uses the same light beams, which for her have equations $t_o - x_o = T_o$ and $t_o + x_o = T_o'$. Use the principle of relativity, or similar triangles, to show $T_o = kT_e$ and $kT_o' = T_e'$ for some constant depending on their relative velocity. Deduce $t_o^2 - x_o^2 = t_e^2 - x_e^2$ for E, and hence the invariance of the interval.

19 Paradoxes of special relativity

The 'Paradoxes'

THE RAILWAY PARADOX
You are seated by the side of the track at a point where two lengths of rail meet. To allow for their expansion when the temperature rises there is a small gap between the two rails. A supertrain approaches, travelling at near the speed of light. You argue as follows.

(i) The driver sees the gap in the rails approach him at nearly the velocity of light. Therefore its length appears greatly contracted and he should enjoy a virtually bump-free ride.

(ii) You see the train approach so fast that its length is greatly contracted. In fact, its length is shorter than the gap and it should vanish entirely into the gap.

Evidently you cannot have it both ways. What does happen?

THE TWIN PARADOX
Peter and Paul are identical twins. Peter stays at home and Paul travels on a space ship to a distant star and back. He attains nearly the speed of light almost all the way, and so Peter sees his twin severely time dilated. Indeed Peter is 70 years older than he was when Paul returns, but Paul is apparently only 4 years older. Or is he? Why discriminate between the twins? What does Paul think of Peter? Paul sees him rush away and return at the same enormous speeds, and so expects a youthful twin to greet him, now somewhat elderly, at the space station. Who is right?

THE SCISSORS PARADOX

(i) A searchlight revolves at a constant angular velocity and the beam moves across the target at a speed which increases with the distance the target is from the searchlight.
    If the target is far enough away, the beam will whip across it faster than light. However, we said that nothing can travel faster than light.

(ii) The beam is replaced with a wand, which still, at its tip, is moving faster than light.

(iii) A guillotine blade, or scissor arm, descends obliquely onto a horizontal beam. Although the blade moves slowly, the point of contact of blade and beam moves sideways at a rate which depends on the angle of the blade to the beam. Therefore it can be made to move faster than light.

(iv) A ball bearing is placed at the point of contact and squeezed between blade and beam. Surely it now travels faster than light?

Paradoxes of special relativity

To resolve any of these paradoxes it is best to start with Einstein's own illustration paradox on the nature of simultaneity which we discussed earlier. Three men, A, O', and B, ride on a supertrain at a velocity near that of light: A at the front, O' in the middle, and B at the back. A and B flash a light signal which reaches O' at the same instant that it reaches a fourth man O standing on the track exactly opposite O' at that instant. The question is, who flashed first?

In the opinion of O' A and B are at rest with respect to him and equally far away. They are all three on the train, and he can check on this any time he likes. Therefore as far as O' is concerned A and B, being equally far from him, must flash at the same time.

In the opinion of O the matter is different. The flashes from A and B must have been sent off before they reached him, and at any moment before O' draws level with O A is nearer to O than B is. Therefore B must flash first.

The conclusion we drew from this was that the idea of simultaneity and priority is relative, and depends on the observer. Let us put both sets of observations onto a space–time diagram. O (with axes X, T) records the space paths of B, O', and A on the train as three oblique and parallel lines. On the same diagram the path of O' is interpreted by O' as stationary duration in time, and so as his time axis. This puts his space axis down as shown to ensure that O' and O agree as to the velocity of light. To O', lines parallel to his space axis represents events occurring at the same time.

Finally, put on the diagram rays of light arriving at O the instant that O' is there, and which emanate from each end of the train. The events 'A flashes' and 'B flashes' are shown. To O (with axes X, T) B flashes first, since O's space axis of simultaneous events is horizontal and B is 'below A'. To O', whose space axis is oblique, A and B flash together.

Conversely the observers O and O' can arrange for A and B to flash in such a way that they both see the flashes arrive at the same time. However, their

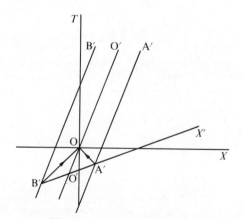

Fig. 19.1. The light flashes used.

192 *Paradoxes of special relativity*

instructions to A and B would differ because one sees A and B at rest, relative to him, and the other sees them in motion.

Guided by this example, let us look at the other paradoxes.

THE RAILWAY PARADOX

Let the gap be AB, let yourself be called O if you will, and the driver O'. He stands at the front of the train which ends at R. As before, you represent what you see on a space–time graph like Fig. 19.2. It does indeed seem that at your time zero O' is well into the gap, and at your time t the whole train O'R is inside the gap where it remains for a short time. Let us now mark on O''s space axis, as before, O'X'. O's axis of simultaneity appears obliquely, and records (3) O' leaving the gap before (2) R enters the gap.

If we had phrased this as a train running through a room we would be in no difficulty and could ascribe the descriptions of ourselves and O' as being due to the relativity of simultaneity. However, a train and a gap suggests something else: the train should fall into the gap. Our description is of an appalling collision; the driver's description is of an almost bump-free ride. What does happen?

The train does fall into the gap for it is pulled into it by gravity, but how far it falls depends on how long it is over the gap. (Remember that vertical distances are not contracted.) Therefore a more detailed analysis would reveal agreement between us and the driver. The train dips into the gap, but not far enough to cause an accident. Given its immense speed horizontally it falls vertically by only a tiny amount.

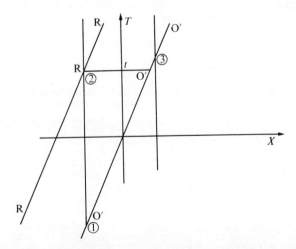

Fig. 19.2. The space–time path of the train: 1, O' enters the gap; 2, R enters the gap; 3, O' leaves the gap; 4, R leaves the gap.

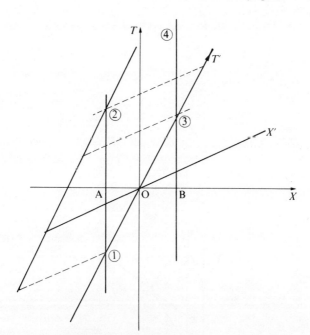

Fig. 19.3. The space–time picture as seen by O' with axes X', T' has the events occurring in the order 1, 3, 2, 4 as shown by the dotted portion of the X' axis, so the train does bridge the gap.

THE TWIN PARADOX

As soon as we draw a space–time diagram the paradox resolves itself. We have drawn Peter's space–time diagram.

Paul does seem to age more slowly than Peter and returns younger as far as Peter sees it. What is happening as far as Paul is concerned? Remember that Paul is accelerating; his path is curved. We can supply him at any instant with an oblique reference frame but at any other instant his reference frames are different. Therefore all our previous simple arguments break down. To see what does happen, take a simple, if implausible, case. Paul travels out at constant speed, and at some point P changes direction instantly and returns, again at constant speed.

At P he switches axes from (1) to (2). During both parts of the journey Peter and Paul see each other age slowly. However, nothing in Peter's life corresponds to Paul's switching axes, and when we look at the diagram we see that Paul has only accounted so far for Peter's life from O to B and C to D. The period from B to C must be added, corresponding to a *change in origin* in Paul's co-ordinates.[1] This correction is the extra age of Peter over Paul

---

[1] This sort of change of co-ordinates is different from those we have met before precisely because acceleration is involved. Paul must ensure that his old and new origins have the same co-ordinates: (0, 0).

194 *Paradoxes of special relativity*

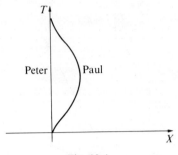

Fig. 19.4.

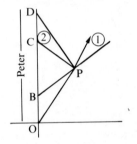

Fig. 19.5. Paul's simplest journey.

when they meet again. In the case of a less drastic acceleration and deceleration, this correction must be made all the time. Therefore we still account for the greater ageing of Peter over Paul.

The situation is not symmetric. Either one twin shifts origin in one fell swoop, or he accelerates and thereby continually shifts origin, and the correction for that is the explanation of the age difference. The important thing is that, even without referring to the rest of the universe, the twins are not symmetrically situated, and we can tell them apart by looking not at relative velocities but at relative accelerations.

THE SCISSORS PARADOX

Yes, the beam of light can travel with a slow angular speed ($\theta°$ per second, say) and yet sweep out a fast-moving image. The image moves at θR units per second, and can indeed move faster than light. However, that image is not a 'thing', it is not an object, it is not the same photons. Once you replace it with a stick the acceleration of the rod from rest to rotation takes time to travel up the rod, and so the rod bends. It bends more and more the longer it is, and the rod never travels faster than light. (In a physically meaningful situation the relativistic mass of the rod and the acceleration of it would also have to be considered.)

The guillotine blade can also move its point of intersection with the beam at any speed. However, it cannot move a physical object along faster than light for the same reason. The inertia and acceleration of the bearing would simply buckle both blade and beam.

You should persuade yourself that the beam of the searchlight transmits signals (the light) but that its tip is not capable of being a signal. Imagine two men on the moon, at the north and south poles, and show that they cannot communicate with such a beam at a rate faster than light.

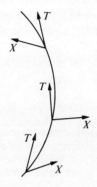

Fig. 19.6. Paul's reference frames change.

20 Gravitation and non-Euclidean geometry

The theory of special relativity with which we have been concerned deals with uniform motion of observers, and with the problems of how it is to be described. The principle of relativity asserts an equivalence between all intertial observers; they should all describe the world in essentially the same way. This theory is subordinate to a greater enquiry which would discuss how the descriptions of non-intertial observers, observers subject to forces and accelerations, differ.

Einstein was interested early on in this problem, which he saw as extending Newton's mechanics. In 1919 he wrote:

When, in the year 1907, I was working on a summary essay concerning the special theory of relativity for the *Jahrbuch für Radioktivität und Electronik*, I had to try to modify Newton's theory of gravitation in such a way that it would fit into the theory [of relativity]. Attempts in this direction showed the possibility of carrying out this enterprise, but they did not satisfy me because they had to be satisfied by hypotheses without physical basis. At that point, there came to me the happiest thought of my life, in the following form:

'Just as is the case with the electric field produced by electromagnetic induction, the gravitational field has similarly only a relative existence. *For if one considers an observer in free fall, e.g. from the roof of a house, there exists for him during his fall no gravitational field—at least in his immediate vicinity.*' (Italics in original.) (Quoted by Holton 1973, p. 364.)

Notice that Einstein has expressed gravitation as a relative phenomena. The relativity is now between an observer subject to a force and an accelerating observer, but let us look more closely at what he means.

We ordinarily detect gravity in a very simple way: we let things go and watch them fall. However, if Einstein's hypothetical, and hapless, observer dropped something as he slipped it would fall alongside him. It would remain beside him as he fell, relatively at rest. Therefore his attempt to detect gravity would fail and for him gravity, at least nearby, would not exist.

We can see this as, in a sense, analogous to the more familiar relativity of motion. Let us enclose the observer in a box, and let him describe his box to a second observer in the depths of space. Both record the same state of weightlessness, we might say because gravity acts equally on all the objects in the first box, although the first observer and his box seem increasingly agitated to us. Now, as no forces are needed to maintain a body in uniform motion but only to change that motion, we can only measure forces by their effects, by the changes they produce. Locally at least, no changes means no forces.

How exact is this equivalence, and what is meant by Einstein's insistence that 'at least in his immediate vicinity' gravitation does not exist for the freely falling observer? If he looks closely at two objects falling along with him, shall we say two balls B_1 and B_2, he will be able to notice a small but increasing relative motion between them. In the first case the two balls are imagined side by side. They both fall towards the centre of the earth, at which point they would collide, and so, given sensitive enough equipment our observer should be able to detect a relative motion of B_1 and B_2 towards each other. Of course, the distance we are from the centre of the earth is so great that this lateral motion is scarcely detectable, and so we say that objects fall in the same direction vertically but we do not make that mistake if they start off far enough apart.

In the second case B_1 is slightly ahead of B_2. Since it is slightly nearer the centre of the earth, the pull on it is greater and it is accelerated down slightly more than B_2. Our observer could observe B_1 drawing slowly away from B_2, corresponding to an increase in the earth's gravitational field. However, this relative motion is again very small, and if we agree to neglect it as too small then we are saying that B_1 and B_2 do remain always equidistant. This amounts to the useful physical fiction of an infinitely distinct centre of an earth of infinite mass, a fiction perfectly adequate for all elementary mechanics. Einstein's equivalence is between a uniformly accelerated system and a system subject to a perfectly uniform force.

However, if we find that the relative motion of the balls is detectable then we are saying that the gravitational field is not uniform, and then it follows that the gravitational field is detectable locally. It is said to be detectable because of its 'tidal' effect, the relative motion of lots of B's (the water molecules in the oceans) giving rise to the tides in response to the varying strength of the moon's gravitation on different parts of the earth. If we imagine B_1 and B_2 to be horizontal and joined by a thin rod, the local effect of the gravitational field would be to bring B_1 and B_2 together and squash the rod. With B_1 and B_2 vertical the effect would be to stretch the rod, and an initially square configuration ABCD would find itself squashed sideways and stretched lengthways as it fell.

Just as the effects predicted by special relativity are only detectable at high speeds, which accounts in part for their strangeness, so the theory of gravitation we are about to describe is only novel for strong fields. We are used to the idea that a falling rod keeps its length, and so it seems if gravity is weak and our measuring equipment poor. However, a powerful, changing field would buckle it, just as the moon bends and manipulates the ocean into tides.

Let us now look at how we can introduce gravitational fields into the space–time world of special relativity.

The conventional element in measurement

Let us imagine intelligent creatures living upon a surface, a two-dimensional world. They wish, naturally, to survey their world, and to do so they can of course only travel upon the surface and measure it, exactly as we survey ours.

198 *Gravitation and non-Euclidean geometry*

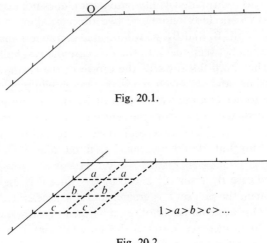

Fig. 20.1.

Fig. 20.2.

$1 > a > b > c > \ldots$

We shall watch them as they make the attempt and record it on the map that they hope to construct.

One point is fixed as the origin for the survey and called O. In one direction a straight line is taken, determined perhaps by the path of a ray of light. We can interpret it as a geodesic if we wish. A second path is then chosen through O, straight like the first but at right angles to it at O. There are two pictures of what is going on, ours and theirs, and theirs might well look like Fig. 20.1. We shall call the lines the co-ordinate axes, and they draw lines looking straight to them. To make the survey they first calibrate their axes, say in kilometres, by successively laying off lengths of 1 km along each axis. They can then survey at least a region near to O like this. From each kilometre marking on the first axis, set off in a straight line at right angles to that axis. As each kilometre mark is passed along that line, measure the distance to the kilometre mark to the right. We can well imagine that they would hope that that distance would be a kilometre too, since then their world would be flat. However, it may turn out to be otherwise. As we see them the measured distances may well diminish for instance, or in general do anything. Let us suppose that they diminish steadily as they move away from O in the regular pattern of the picture.

What sense can they make of their map? The straight lines alongside the (2) axis appear to be converging. If they decided to abandon flatness they could redraw their paths in a more visually suggestive way, the new, curved appearance of the paths still suggesting geodesics, of course, in their new picture of their world. We can literally see them upon the curved surface of a sphere in our three-dimensional world, a description which they could take metaphorically, as a figure of speech. They could consider themselves to be representing their curved world on a curved map as shown in Fig. 20.3.

If they prefer, as we do, to have flat maps they can still make their geodesic paths look straight, but only by making distances unintuitive. By continually varying the scale of their map, and therefore where they represent the position of the kilometre posts, the paths on the map change their shape and spacing. By spacing them further and further on the map they can succeed in drawing a map where the geodesic paths look straight but seem to be ever more spread out. This projection we would call geodetic, for the reason that geodesics on the sphere are drawn as geodesics on the map.

Therefore we can imagine them with three natural interpretations, at least, of the world they live in. It might seem to them curved, or flat but with convergent straight lines perpendicular to the axes, or flat but with a distorted distance. They cannot rule out any of these pictures as wrong, since evidently they are equivalent. How might they defend a choice of either the second or third, which might please them more as being free of reference to a third dimension? In the second case a kilometre reading from O to X was measured as less than 1 km between YW. A rival description would be that any object 1 km long at O, like the measuring rod, stretches on travelling outwards to Y. Some strange force expands it as it travels, so that when it arrives at Y it exceeds YW and so YW appears less than 1 km. Consider for instance 1 km of taut rope, which will lie between O and X exactly but will surpass YW.

In the same way the geodetic map can be re-interpreted as literally accurate in a world subject to a universal force which expands objects as they move outwards. On such a map using a rope to lay off a curve of equal distance to the (2) axis produces the figure opposite.

To us such a universal force is displeasing, acting equally and without hindrance upon all objects and seemingly without cause. It is *ad hoc*; its only purpose is to restore the world to flatness. However, we cannot refute the creatures in the surface if they wish to believe it. A delightful example of Reichenbach's (1958, p. 11) may help to emphasize this point.

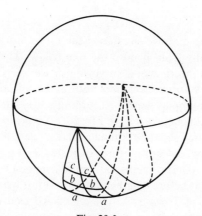

Fig. 20.3.

200 *Gravitation and non-Euclidean geometry*

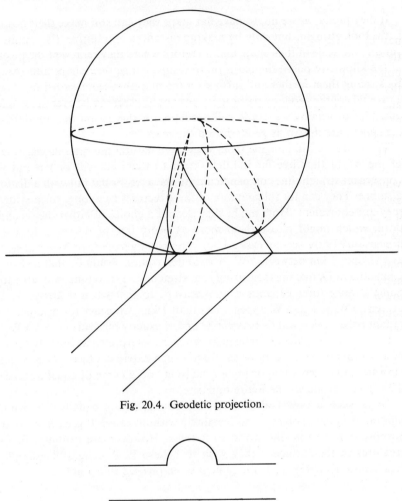

Fig. 20.4. Geodetic projection.

Fig. 20.5.

Two worlds rest one above the other; the upper world is transparent and shadows fall vertically from it onto the one below. The beings in the lower world accept the calibration and surveying of the ones above, but differ in interpretation. Away from the hump both agree, of course, and the upper beings perceive the hump as we do. Accordingly they survey it as in Fig. 20.6. However, the lower beings see it as in Fig. 20.7. Entering or leaving the region beneath the hump they see their kilometre posts seemingly disturbed (to us). To them, it seems that a 1 km object is distorted as it passes through BA. As it travels from right to left it seems to them to lengthen markedly at first, then settle down somewhat before lengthening against just before returning quickly to its original length as T passes A. The region AB seems to them to lengthen objects that pass through it.

Gravitation and non-Euclidean geometry 201

We can only conclude that any act of surveying has an element of arbitrariness or conventionality about it. We cannot be sure but can only assume that the measuring rods we carry around with us do not change their lengths from point to point in some universal way. We may adopt as a convention either that rods remain constant in length or that they vary under some strange force, and our choice can be made, for instance, on grounds of naturalness or ease of computation. In the first case we would be inclining to the first option, and in the second case possibly towards the second.

There is a delightful interpretation of the spherical universe with which you may be familiar.[1] Suppose that you have a two-dimensional world, a plate, which is heated non-uniformly, the temperature rising steadily as you move away from the centre of the plate. You wish to survey the plate, unaware that it is heated, and to do so you push a metal ruler around on it, which is small by comparison with the plate. As it is moved towards the rim it expands, and it is possible to specify a function relating the temperature at a point on the plate to the distance from that point to the centre so that the following happens. The co-ordinate grid imposed on the plate broadens as it reaches towards the edge, and exactly resembles a geodetic map of a region of the southern hemisphere of a globe. The heated-plate universe is intrinsically the same as the spherical universe.

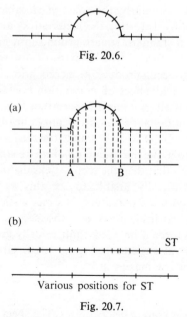

Fig. 20.6.

(a)

A   B

(b)

ST

Various positions for ST

Fig. 20.7.

---

[1] I do not know who first thought of it. It is to be found e.g. in R. P. Feynman, *Lectures in physics*, Vol. III, § 42.1 (1966).

202 Gravitation and non-Euclidean geometry

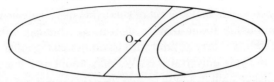

Fig. 20.8. The heated-plate universe: three equidistant curves.

The function taken relating temperature and radial distance from the centre (extrinsically measured!) is $T = 1 + K^2 r^2$ where K is some constant. If we take $T = 1 - K^2 r^2$ we obtain the *cooled-plate universe*. Now temperature diminishes as we move away from the centre, and the ruler contracts. Equal intervals as measured by this ruler appear to us to crowd together as you go out towards the rim. What does this universe look like?

The key picture is that of a line asymptotic to a co-ordinate axis. The axis is drawn through O and P is a point on the plate. Of all the straight lines through P, some meet the axis and some do not. However, there is one which draws nearer and nearer to the axis but as it does so it also draws nearer to the rim, so that 'distances shrink' in such a way as to balance the two effects. It would meet the axis on the rim, but alas, there the temperature is zero, the ruler of zero length. The rim cannot be reached; the line and the axis are asymptotic. The cooled-plate universe is, as perhaps you guessed, the non-Euclidean plane.

Unlike the plate universes and sphere universes Reichenbach's humped universe is not uniform. Similarly, the effect of variable gravity is to buckle the squares of the grid by distorting the meter sticks and so also the clocks. Any attempt to carry a standard square around in a gravitational field will subject that square to forces which, extrinsically seen, will buckle the square or, intrinsically seen, produce a curved grid. The observed effects corresponding to the tidal effect of gravitation buckle and curve the grid. Gravitation manifests itself as a curved space-time[2].

To observe the gravitational field in his vicinity our observer, above, would have to see the square ABCD squashed and stretched as he moves it around. However, if he chose to make the square up of metre sticks and clocks and to measure everything with that then the metre sticks distort. He is like the plate measurer who is intrinsically unable to see this as distortion, although intrinsically able to record the curvature. His metre sticks seem buckled and stretched only to an extrinsic observer, who sees a flat space–time and interprets gravity as a force if he wants; intrinsically gravitation is absorbed directly into the metrical structure of space itself. Gravity is the shape of space as determined by the masses in it.

Example

Upon a taut elastic sheet place a heavy object. The sheet sags a little under the weight, and if we roll a marble over the sheet it does not roll straight by but

---

[2] In special relativity, space remains Euclidean.

dips in slightly towards the object as it passes. We could say that the marble was attracted by the gravitational pull of the object. Now, keeping the sheet exactly where it is, remove the object. The marble will roll exactly as before, and so we could say that the curved surface of the sheet expresses the pull of the object on the marble at each point in its path.[2]

There is therefore a gravitational effect on the path of a ray of light, which reflects the 'curved' nature of geodesics on a curved surface. The geometry obtained, which reflects the gravitational effect, is determined intrinsically. The trumpet-shaped surface is negatively curved—in the intrinsic metric—try drawing a square on it.

Exercises

20.1 You might care to adopt a suggestion of Helmholtz and transplant the usual co-ordinate description of space to the image of space in a spherical mirror, curved inward or outward. The worlds in the mirror are in some ways like the non-Euclidean ones. Which is which? Details are in Helmholtz's delightful essay *On the origin and significance of geometrical axioms*, available in *The world of mathematics* (1960, Vol. 1, pp. 647–68. Allen and Unwin). Other essays in the same volume of equal relevance and charm are by Clifford.

Helmholtz also points out that this world, seen through a lens of negative focal length, appears non-Euclidean. Consider the opposite case, in which the real world is non-Euclidean but our eyes are adapted by lenses to see it as Euclidean, and thereby convince yourself of the plausibility of non-Euclidean geometry as a true geometry of space.

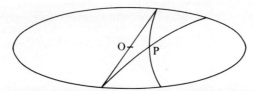

Fig. 20.9. The cooled-plate universe: two pairs of asymptotic lines.

---

[2] In fact the calculations on this model give half the correct result, the missing half being a special relativistic effect.

21 Speculations

Up to now we have maintained a bold pretence that space–time can be thought of as two-dimensional, as a surface, and as such is amenable to the theorems of differential geometry that we have described. Since space–time is, of course, really four-dimensional we must consider how our descriptions break down. In fact they do not fail us badly. Just as we are happy doing Euclidean geometry in a plane and arguing by analogy about the properties of figures in three dimensions (seeing them for instance as made up out of plane slices) so the geometry of space–time is largely captured in its two-dimensional aspects. Precisely because we have only an intrinsic view of it anyway the gain in intuitive understanding probably outweights the capacity for analogies to mislead. However, in four dimensions it ceases to be true that '4-manifolds' are so easily characterized by their curvature. For a start, curvature is no longer one number, which can vary from point to point, but many (the details fortunately need not concern us) and we cannot describe surfaces so simply in terms of their curvature. However, once this is admitted, as it were in the small print, we can proceed as before. Three-dimensional pictures are even better, and the analogy of masses depressing a rubber surface with gravity is a good one. The curvature induced in the rubber sheet is a fair picture of how nearby masses are attracted by 'gravity' to the central mass. From now on we shall try and draw three-dimensional pictures to suggest the space–time situation we have in mind.

Notice, however, that length contraction takes place only in the line of flight anyway, and so affects x distances but not y or z distances, so to speak.

Gravitation

The mass of a star distorts the space around it, curving it, and this is detected by the effect it has on rays of light. Light seeks out not magically 'straight' paths, but the shortest possible path. These geodesic paths, curved because the space is curved, are what reveals the gravitational effect of the star. The celebrated observations of this effect on light emitted from a star, conveniently situated behind the sun during the eclipse of 1919, were a novel consequence of the general theory of relativity and helped it to gain acceptance, although the numerical fit was not good (see Kilmister 1973, d'Abro 1927 (reprinted 1950)). Operationally, of course, the sequence of observations is this: clocks and rods determine the metric, and with respect to that metric the path of light is geodesic. However, computation reveals the space on that metric to be curved; 'gravity', the geometry of the light rays, is not Euclidean. It would be possible to describe the space in Euclidean terms, but then light would travel in a curved (non-geodesic) fashion and physics would lose its simplicity.[1] Whether or not the space is strictly non-Euclidean (homogeneous) or curved in varying ways we shall consider later.

Notice that our programme had been to work locally, determining the metric locally and from that the local effect of gravity. Patching all these local views together gives a global view, with global properties such as the bent path of light, but action at a distance has been replaced. The force of gravity no longer reaches across neutral space to pull in objects, the objects instead feel the local curvature of space that gravity 'is'. Just as in the simple Newtonian case, the mathematician still aims to describe how this curvature is determined by the distribution of masses in space. However, the problem is much harder now, for the masses directly change the metric which they left alone in the Euclidean–Newtonian theory. The equations Einstein produced, the field equations, express the problem mathematically but do not solve it. The simplest solution produced, that due to Schwarzschild, was for a static spherically symmetric body (like a star could be) and its predictions differed only slightly from Newtonian ones. The perceptible differences, however, were in its favour and the new theory gained in credibility as a result.

Happily the theory has some descriptive aspects which we need no detailed mathematics to explain. Imagine you are sitting in a spaceship, surveying the space around a star. For your geometry you take the geometry of light rays which are emitted by small beacons situated at various places in the orbit.

Since the effect of the gravitational field is to try to pull the light inwards, if you seek to make up a square of light rays you must aim a little outwards with your beam of light. If the gravitational field is imagined as a distorted rubber sheet with the star at its centre, then the task is to define the shape of

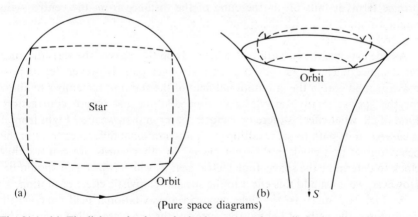

Fig. 21.1. (a) The light paths shown by broken curves do not lie in the plane of the orbit. (b) The angles of the dotted square are less than $\pi/2$—why?

---

[1] For the origin of these ideas see H. Poincaré, *Experiment and geometry*, in *Science and hypothesis* (1905, reprinted 1952). Pergamon, Oxford.

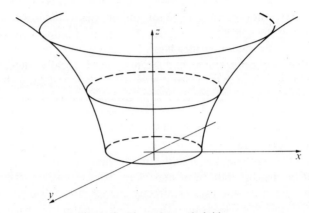

Fig. 21.2. Flamm's paraboloid.

this surface in such a way that geodesics upon it represent rays of light. Intuitively one feels that the sides of the well formed by the sheet should draw together and become increasingly steep as one nears the centre (the star) in order to represent the increasing force of gravity. The surface which most accurately captures this is the trumpet-shaped space shown in Fig. 21.2, which is obtained by rotating the top half of the parabola $z^2 = 8m(y-2m)$ in the y, z plane about the z axis and is known as Flamm's paraboloid. In the intrinsic geometry of this surface the curvature is everywhere negative and decreases as you move away from the centre. Geodesics on it are obtained by, so to speak, aiming a little outwards and rolling back in, as a marble would. The varying curvature provides for the local or tidal effect of gravity, and in precise terms it falls off as the cube of the distance from the centre—you might compare this with the pseudo-spherical model of a space of constant negative curvature.

As one expects, the curvature also depends on the mass of the star; distance for distance bigger stars should exert a stronger pull. However, let us now move around within the gravitational field of the star and remember to allow for the special relativistic effects of time dilation and length contraction. First of all, what effect is there on a clock nearer in than we are? Light from it is curved on its path to us, resulting in an observational difference to us in the behaviour of that clock in different places. In other words, we can use that clock to determine the space–time metric near it, which is just as it should be. However, we must add this effect to the special relativistic effect of motion.

As Fig. 21.3 suggests, the light cones in a gravitational field vary. Their asymptotes, the paths of light at the vertex of the cone, are pulled over by the gravitational effect. However, if we choose a region so small that the gravitational effect is constant (no tidal effect) the light will seem to travel with its usual constant velocity. Therefore we can still infer locally what the pure time and pure space axes are at each point in a gravitational field. To put it another way, over small regions of space–time the effects of gravity can be ignored and space–time is locally flat with familiar properties. A more accurate experiment would reveal space–time to be curved; this is measured in the change in orientation of the light cones from point to point. This is depicted in Fig. 21.4. Notice that we still do have perfectly good space and time axes everywhere. It is possible to imagine colossal curvatures, which would disturb the light cones considerably.

Black holes

You have detected a curved region of space (and space–time) and can see nothing at its centre. Cautiously you map out the light cones, and it turns out that they look like Fig. 21.4. At a certain distance from the invisible centre

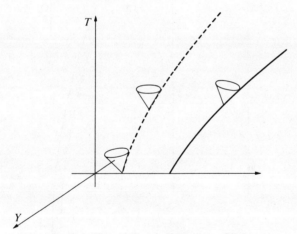

Fig. 21.3. Dotted lines indicate rays of light.

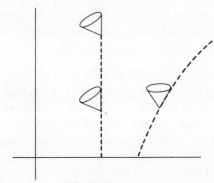

Fig. 21.4.

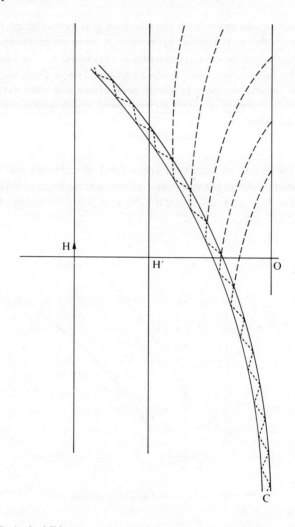

Fig. 21.5. An invisible mass at H attracts the clock C which falls under gravity as shown. It radiates out to us at O, but its pulses seem dilated and we never see it enter the critical region HH′ although it does so in finite proper time.

they have one asymptote vertical. In terms of the path of light rays it means that light at that point travels vertically, i.e. remains at that distance from the centre of the region. As for the paths of rays of light they look like Fig. 21.5. Any event inside the critical distance cannot radiate outside because the light cone there must point wholly inwards, which is why the centre is invisible, and such a thing would be called a black hole.

Suppose you now release a clock. It falls towards the black hole, gathering speed as it goes. Its path is shown in Fig. 21.5. Since it must always travel slower than light it does not travel along a light ray path but along a gentler curve. The clock continues to beat and enters the critical region in finite proper time. However, the signals it sends us take longer and longer to return, as Fig. 21.5 also suggests, because they must fight against steadily greater curvature. We in fact see the clock take an infinite amount of our time to reach the critical region (an exaggerated Peter and Paul effect.)

Therefore the boundary of the critical region is not something that hits you. You do not notice it as you enter, nor do we notice you enter. However, once inside you cannot escape. If light cannot escape, and you must travel slower than light, you are trapped. Indeed, anything in that region must continue to fall inwards, experiencing steadily greater and greater forces until, even at the atomic level, it is ripped apart. A black hole is a region of immense gravitational power, curving space around a central core from which there is no escape.

Sometimes it is objected that if not even light can escape then nothing can, so how can it still tell us that it is there? Imagine a visible star which somehow rapidly increases in mass. At some stage the critical region envelopes the star, once the mass is sufficiently great, and it becomes invisible. Certainly until then it exerted its gravitational field on space around it. If at the moment of invisibility it switched off that field that would be a signal, travelling out with at most the speed of light. However, what a black hole cannot do is to radiate information, and so its very blackness prevents it from removing its gravitational influence.

There is nothing in this description of black holes which could flout physical theory, unless it be the hypothesis that such extremes of curvature are possible. They do not extend the range of forces much beyond the enormous influence of gravity on neutron stars, which are observationally well established, and indeed there is some evidence that they actually exist. Cygnus X-1, a powerful radio source, is thought by some to be near a likely candidate.[2] However, they do raise unfortunate consequences for the theory, for at the centre of a black hole is a singularity in the space–time, a region where the laws of physics cease to apply. As Professor Roger Penrose (1973, p. 121)[3] puts it:

... it is hard to avoid the inference that tidal effects which approach infinity will occur, producing a region of space–time where infinitely strong gravitational forces literally squeeze matter and photons out of existence.

If such a region could be formed which would still be capable (unlike black holes) of emitting signals, the 'naked' singularity, as it is then called, could do what it likes. As far as physics is at present formulated there is nothing

---

[2] The hole might be some 60 miles across, in orbit around a giant star about 18 000 000 miles in diameter!
[3] Much of the material in this section is taken from Penrose (1973).

governing what it could or could not do, and theory does allow such things to be formed out of large bodies stirred up into a certain rotation. The dilemma for cosmologists is real at all events, whatever the observational status of the processes they have conjured up. Usually, as Penrose points out, a singularity in a theory heralds the breakdown of that theory unless it can be recast in a new form. When that theory is a theory of space and time the problem is all the more severe.

It may be that the way out will take us back towards problems raised by the Ancient Greeks. Although their ideal of one natural enquiry into everything has broken down, we still wrestle with an interplay between mathematics and the physical sciences. This interplay rests in part on a mathematization of a process which includes the hypothesis that space and time are smooth, continuous-like surfaces, rippled and curved perhaps but not broken and disjointed. By contrast matter is indivisible and atomic. It would be desirable, some argue, to analyse matter in terms that resembled those we use to analyse space, to describe it using the logically necessary concepts of physics much as the fundamental particles displace tables and chairs. The reality of matter is not in question; what is is the nature of matter. If a portion of space is pinched to a point, the distortion produced could perhaps resemble a gravitational field, in which case space would consist of nothing but space, clumps of it forming matter and spreads of it forming gravity. However, when we consider the changes in curvature involved it seems that we must, at present, consider infinite curvatures. It might be preferable if we could reformulate this interplay between matter and space without such vast curvatures occurring. Remembering that, in two dimensions at least, positive curvature increases as the radius of curvature decreases (and analogously for negative curvature) we see that to prescribe the maximum amount of curvature allowed is to prescribe a lower limit to the size of clumps of space. There would be a meaning to the size of the clump; it would be the size of the particle it represented. A reason to limit the minimum allowed size of a particle would be provided by quantum-mechanical considerations. You could argue that the energy needed to measure such distances was absurdly large. Therefore a limit to curvature is theoretically tempting, but in that case the idea of space as a surface must go, since it cannot after all be infinitely distorted. In asking the question 'what is space?' do we not hear an echo of the Greek concern for what a line is, and how to relate the concepts of line and point? Or is it a fancy of our own?

These speculations from cosmology, with their hint of links between the very big and the very small, between 'fact' and 'explanation', are as far as we can go, yet a surprisingly early and charming exposition of it was written by W. K. Clifford in 1876 and survives in abstract. Let me remark upon two points only: the relationship between matter and curvature broached in (3), and the doubt about continuity expressed in (4). One could not wish for a more exciting programme of research.

Appendix

On the space theory of matter by William Kingdon Clifford (abstract)

Riemann has shown that as there are different kinds of lines and surfaces, so there are different kinds of space of three dimensions; and that we can only find out by experience to which of these kinds of space in which we live belongs. In particular, the axioms of plane geometry are true within the limits of experiment on the surface of a sheet of paper, and yet we know that the sheet is really covered with a number of small ridges and furrows, upon which (the total curvature not being zero) these axioms are not true. Similarly, he says although the axioms of solid geometry are true within the limits of experiment for finite portions of our space, yet we have no reason to conclude that they are true for very small portions; and if any help can be got thereby for the explanation of physical phenomena, we may have reason to conclude that they are not true for very small portions of space.

I wish here to indicate a manner in which these speculations may be applied to the investigation of physical phenomena. I hold in fact

(1) That small portions of space *are* in fact of a nature analogous to little hills on a surface which is on the average flat; namely, that the ordinary laws of geometry are not valid in them.
(2) That this property of being curved or distorted is continually being passed on from one portion of space to another after the manner of a wave.
(3) That this variation of the curvature of space is what really happens in that phenomenon which we call the *motion of matter*, whether ponderable or etherial.
(4) That in the physical world nothing else takes place but this variation, subject (possibly) to the law of continuity.

I am endeavouring in a general way to explain the laws of double refraction on this hypothesis, but have not yet arrived at any results sufficiently decisive to be communicated.

22 Some last thoughts

The vital point is that we now recognize the arbitrary nature of our assumptions. The complicated assumptions fundamental to the theory of relativity may be shocking to the layman, if he be honest enough to acknowledge that he does not understand them, but they no longer shock the mathematician. The point which I wish to insist on in this connexion is that it is to the doubts about Euclid's parallel postulate, and efforts of such thinkers as Saccheri, Lobachevskii, Bolyai, Beltrami, Riemann, and Pasch to settle these doubts, that we owe the whole modern abstract conception of mathematical science. (Coolidge 1940, (reprinted 1963), p. 87.)

Meanings

In various ways at various times people have tried to give meaning to the world around them, and give rational tongue to the wonder it evokes. The heroic Greek attempt to reach deductive understanding survived longest in the Euclidean description of space. The extension of mathematics to embrace non-Euclidean geometries seemingly ended that naïve attempt, but left a possibility that at least mathematics itself could be incontrovertibly grounded in logic. That attempt failed as well, and it would seem at present that we can have no unguarded commitment to either our mathematics or our physics. Yet parallel to that increasing uncertainty in our intellectual enquiry has come a corresponding growth in the richness and power of our descriptions. The tools we have no longer answer the original questions, but answer successfully others no less useful. They are discursive fictions. The understanding this brings and the meanings this confers would seem to be part of man's creation. Here, as elsewhere, we make the future, but not in circumstances of our own choosing.

Last mathematical appendix

If the sphere has radius R and curvature $1/R^2$, points on it have extrinsic co-ordinates (x, y, z) where $x^2+y^2+z^2 = R^2$. The instrinsic co-ordinates of P are (u, v) with axes as shown and

$$x = \frac{uR}{(1+u^2+v^2)^{1/2}} \quad y = \frac{vR}{(1+u^2+v^2)^{1/2}} \quad z = \frac{R}{(1+u^2+v^2)^{1/2}}.$$

The co-ordinates are longitudes, e.g. $u = 0$, and great circles perpendicular to $u = 0$. The distance

$$ds^2 = dx^2 + dy^2 + dz^2$$

becomes
$$R^2 \left\{ \frac{(1+v^2)\,du^2 - 2uv\,du\,dv + (1+u^2)\,dv^2}{(1+u^2+v^2)^2} \right\}.$$

Writing $ds^2 = dx^2 + dy^2 + dz^2$ does not imply tunnelling into the sphere, since we insist on $x^2 + y^2 + z^2 = R^2$ always.

The formidable expression for ds^2 has this meaning. If we consider $v = 0$ always (the u axis) and so $dv = 0$ also, we move out along the u-axis:
$$ds^2 = \frac{R^2 du^2}{(1+u^2)^2} \quad \text{i.e.} \quad ds = \frac{R\,du}{1+u^2}$$

Now the u axis sits upon the sphere above the x axis, and x cannot exceed R. Along the u axis we have $x = uR/(1+u^2)^{1/2}$ and as x gets nearer and nearer 1 we find $u \to \infty$. Therefore although the x, y co-ordinates of the sphere are always between -1 and $+1$, the u, v co-ordinates are without limit. Increasing u steadily by a fixed amount du is therefore possible; we cannot reach an 'edge'. However, the increase in ds, the distance from the north pole, falls off as u increases since[1] $ds = R\,du/(1+u^2)$. Thus the constant step results in a diminished move away from the origin. In other words, the measuring rod increases as you move away from the origin by an amount proportional to $1+u^2$.

Now following Beltrami we consider a surface of curvature $-1/R^2$, which we can do essentially by writing iu for u and iv for v. Then
$$ds^2 = R^2 \left\{ \frac{(1-v^2)\,du^2 + 2uv\,du\,dv + (1-u^2)\,dv^2}{(1-u^2-v^2)^2} \right\}.$$

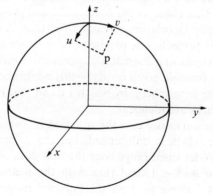

Fig. 22.1.

---

[1] Indeed a simple integration shows the maximum value of s is
$$s = \frac{\pi R}{2} = R \int_0^\infty \frac{du}{1+u^2}$$

(By considering the arc length along a curve making an angle θ with $v = 0$ we can obtain the trigonometric formulae of Bolyai and Lobachevskii. We shall not do this.) We have

$$x^2 + y^2 + z^2 = \left\{\frac{iuR}{(1-u^2-v^2)^{1/2}}\right\}^2 + \left\{\frac{ivR}{(1-u^2-v^2)^{1/2}}\right\}^2$$

$$+ \left\{\frac{R}{(1-u^2-v^2)^{1/2}}\right\}^2$$

$$= \frac{(-u^2-v^2+1)R^2}{1-u^2-v^2} = R^2,$$

if one writes $x = uiR/\sqrt{(1-u^2-v^2)}$ etc. corresponding to the eighteenth century intuition of a sphere of imaginary radius.

If we once again set $v = 0$ and therefore $dv = 0$, the formula for arc length becomes

$$ds = \frac{R\,du}{1-u^2}$$

along the curve $v = 0$. As one moves away from the point $(0, 0)$ in fixed steps of du the amount ds increases, and contrariwise for a fixed ds du must increase as u does. Since it would be ds that creatures with this geometry would consider distance we see that their geometry is that of the cooled-plate universe. It should be pointed out that in this case u and v are bounded, indeed $u^2 + v^2 < 1$.

It is helpful to consider both these descriptions as map projections. In the positive curvature case u and v were unbounded, so we let them range over a plane. Think of that plane as the north pole, and project radially inwards to the centre, thereby associating to each point (u, v) a point on the northern hemisphere. This projection is precisely the map we have already discussed. We now define (following Riemann's approach) a metric on the (u, v) plane by means of the formula given for ds^2 — this establishes an isometry between the plane and the hemisphere, by which I mean that the intrinsic geometries on the two surfaces are identical.

In the negative curvature case the situation is reversed. Now it is the non-Euclidean space which is unbounded, but the co-ordinates in it, u and v, are bounded. We let them range over the unit disc, or more strictly over its interior where $u^2 + v^2 < 1$, and then with the metric given by the second expression for ds^2 above we have the Poincaré model of the non-Euclidean plane, which we can think of as our map of it. Naturally when we have a map we look for the original territory, but this time we fail to find one because of the frequently mentioned results of Hilbert's that no isometric embedding of the non-Euclidean plane into our three-space exists. In a sense the map is the best picture of it we can have.

Exercise

22.1 (Very hard). Show that the formula for arc length along a horocycle reduces to an expression which shows that the induced geometry upon it is Euclidean, i.e. as a line, the measure of length upon it obtained from the non-Euclidean metric in the disc is Euclidean. Hint: The horocycle can be taken to be $(u-a)^2+v^2 = r^2$, and since this circle must touch the boundary circle at $(1, 0)$ $(1-a)^2 = r^2$; furthermore $(u-a)\,du + v\,dv = 0$. Substitute into the expression for ds^2 and simplify. This also shows that the induced geometry on the horosphere is Euclidean. Can you realize this isometry as a simple projection in the way done above for the positive curvature case?

There is one last point to be made, which, although it seems at first to discuss the similarities between non-Euclidean geometry and special relativity, turns out to bear on the question of isometric embeddings.

In the *Erlanger Programm* of 1872 Felix Klein argued that in any geometric situation one should always consider the group of transformations which leave invariant the geometric properties you are interested in. Thus in the case of the usual plane geometry the allowed transformations must preserve distance, x^2+y^2. Accordingly they are translations, rotations, reflections, and combinations of these. If we exclude reflections, which necessarily reverse the orientation of our figures, we are still left with a group of transformations, and of these the non-identity translations move every point while the rotations each have a fixed point, the centre of the rotation. If we fix a point arbitrarily and call it the origin we can then consider the group of all rotations about the origin. It is called the special orthogonal group, SO(2), and can be represented faithfully by the group of all matrices of the form

$$\begin{pmatrix} \cos\theta & -\sin\theta \\ \sin\theta & \cos\theta \end{pmatrix}$$

where θ is the angle of the rotation.

The analogous group in three dimensions is, of course, called SO(3) and it consists of all matrices of the form

$$\begin{pmatrix} \cos\theta & -\sin\theta & 0 \\ \sin\theta & \cos\theta & 0 \\ 0 & 0 & 1 \end{pmatrix} \times \begin{pmatrix} \cos\phi & 0 & -\sin\phi \\ 0 & 1 & 0 \\ \sin\phi & 0 & \cos\phi \end{pmatrix} \times \begin{pmatrix} 1 & 0 & 0 \\ 0 & \cos\psi & -\sin\psi \\ 0 & \sin\psi & \cos\psi \end{pmatrix}$$

where θ, ϕ, ψ are the so-called Euler angles. Since SO(3) has been set up to preserve $x^2+y^2+z^2$ and to fix the origin it is simply rotations of the sphere; it therefore has a subgroup of rotations fixing the north and south poles which can easily be seen to be isomorphic to SO(2), i.e. those matrices of the form

$$\begin{pmatrix} \cos\theta & -\sin\theta & 0 \\ \sin\theta & \cos\theta & 0 \\ 0 & 0 & 1 \end{pmatrix}.$$

In the case of two-dimensional Lorentz transformations with an agreed origin the preserved quantity was $x^2 - t^2$ and we shall refer to the group of all such transformations as SO(1, 1). It exists inside a larger group called SO(2, 1), which consists of all transformations about the origin preserving $x^2 + y^2 - t^2$. Just as SO(2) and SO(3) were realized as rotations, so SO(1, 1) and SO(2, 1) can be realized as hyperbolic motions. Recall that any Lorentz transformation in two variables sent a point on some hyperbola to another point on the same hyperbola. In three dimensions we can consider, without loss of generality, the locus $x^2 + y^2 - t^2 = -1$, i.e. all events whose time-like separation from the origin is one unit. These form a two-sheeted hyperboloid, i.e. two symmetrical bowls, one lying inside the forward light cone of the origin and one lying inside the backward light cone. Fix one, say the forward one, in your mind. Consider the subgroup of SO(2, 1) fixing the point (0, 0, 1) in this sheet; once again it is not too difficult to see that it is isomorphic to SO(2) which is thought of this time as rotations of the bowl about the t axis.

In summary, in the first case we have SO(3) acting on a sphere with SO(2) as the subgroup that fixes a point, and in the second case we have SO(2,1) acting on a bowl with SO(2) as the subgroup that fixes a point. Now there is a metric given in the (x, y, t) space, namely $ds^2 = dx^2 + dy^2 - dt^2$. We state, but do not prove, that this induces a metric on the bowl with respect to which the bowl becomes a model of non-Euclidean space. Indeed, if we take (u, v) co-ordinates on the bowl, where $x = \sinh u \sin v$, $y = \sinh u \cos v$, and $z = \cosh u$, then[2] $ds^2 = du^2 + \sinh^2 u\, dv^2$. Therefore we do indeed have an isometric embedding of the non-Euclidean plane, not into Euclidean three-space but into relativistic three-space. This ultimately is the explanation for the very close tie-up between the co-ordinate transformations of non-Euclidean geometry and special relativity.

Exercise

22.2 Show that SO(2, 1) has a subgroup consisting of all matrices of the form

$$\begin{pmatrix} \cosh u & 0 & \sinh u \\ 0 & 1 & 0 \\ \sinh u & 0 & \cosh u \end{pmatrix}.$$

What is its effect on a suitable slice through the bowl? Find another similar subgroup.

---

[2] If one sets $x = R\cos\theta \sin\phi$, $y = R\sin\theta \sin\phi$, $z = R\cos\phi$ on the sphere $x^2 + y^2 + z^2 = R^2$ then the $ds^2 = R^2(\sin^2\phi\, d\theta^2 + d\phi^2)$ and the curvature works out to be $1/R^2$.

List of mathematicians and physicists

1717–1783	D'Alembert, Jean Baptiste le Rond	1749–1827	Laplace, Pierre Simon
1786–1855	Arago, François	1752–1833	Legendre, Adrien Marie
1835–1900	Beltrami, Eugenio	1832–1903	Lipschitz, Rudolf Otto Sigismund
1802–1860	Bolyai, Janos	1792–1856	Lobachevskii, Nikolai Ivanovitch
1775–1856	Bolyai, Wolfgang Farkás		
1781–1848	Bolzano, Bernhard	1853–1928	Lorentz, Henrik Antoon
1845–1879	Clifford, William Kingdon	1852–1931	Michelson, Albert Abraham
1824–1873	Codazzi, Delfino		
1879–1955	Einstein, Albert	1864–1909	Minkowski, Hermann
1707–1783	Euler, Leonhard	1746–1818	Monge, Gaspard
1768–1830	Fourier, Jean Baptiste Joseph	1860–1937	Morley, Frank
		1843–1930	Pasch, Moritz
1777–1855	Gauss, Carl Friedrich	1788–1867	Poncelet, Jean Victor
1821–1894	Helmholtz, Hermann Ludwig Ferdinand von	1826–1866	Riemann, Georg Friedrich Bernhard
1719–1801	Kaestner, Abraham Gotthelf	1667–1733	Saccheri, Gerolamo
		1780–1859	Schweikart, F. K.
1724–1804	Kant, Immanuel	1794–1874	Taurinus, Franz Adolph
1849–1925	Klein, Christian Felix	1633–1711	Vitale, Giordano
1736–1813	Lagrange, Joseph Louis	1616–1703	Wallis, John

Bibliography

Aaboe, A. (1964). *Episodes from the early history of mathematics.* New Mathematical Library, New York.
d'Abro, A. (1927 (reprinted 1950)). *The evolution of scientific throught.* Dover, New York.
Alexandrov, A. D. (1963). Non-Euclidean geometry. In *Mathematics, its content, method, and meaning*, Vol. 3, Chap. 17, pp. 97–189. MIT Press, Cambridge, Mass.
Beltrami, E. (1868). Saggio. . . . G. Mat. **6**, 248–312.
Biermann, K. R. (1969). *Naturwissenschaften Tech. Med.* **10** (1), 5–22.
Blumenthal, L. M. (1961). *A modern view of geometry*, p. 53. Freeman, San Francisco.
Bolyai, J. (1831). *Science absolute of space* (transl. G. B. Halsted); see Bonola 1912.
Bonola, R. (1912 (reprinted 1955)). *Non-Euclidean geometry* (transl. H. S. Carslaw). Dover, New York.
Boyer, C. B. (1968). *A history of mathematics.* Wiley, New York.
Carruccio, E. (1964). *Mathematics and logic in history and contemporary thought* (transl. I. Quigly). Faber, London.
Cassirer, E. (1950). *The problem of knowledge.* Yale University Press, Newhaven, Conn.
— (1923 (reprinted 1953)). *Substance and function, and Einstein's theory of relativity.* Dover, New York.
Codazzi, D. (1857). Intorno alle superficie. . . . Ann. Sci. Mat. Fis. **8**, 346–55.
Coolidge, J. L. (1940 (reprinted (1963)). *A History of geometrical methods.* Dover, New York.
Coxeter, H. S. M. (1961). *Introduction to geometry.* Wiley, New York.
— (1955). *The real projective plane*, 2nd edn. C.U.P., London.
Daniels, N. (1975). *Thomas Reid's 'Inquiry'.* Burt, Franklin and Co,
— (1975). *ISIS.* **66**, 75–85.
Dicks, D. R. (1970). *Early Greek astronomy to Aristotle.* Thames and Hudson, London.
Dictionary of scientific biography (1970–6). Scribners, New York.
Duhem, P. (1954). *The aim and structure of physical theory* (transl. P. P. Wiener), p. 56. Princeton University Press.
Dunnington, G. W. (1955). *Gauss: titan of science*, p. 177. Hafner, New York.
Einstein, A. (1923 (reprinted 1952)). *The principle of relativity (selected papers by Einstein and others, in translation).* Dover, New York.
— (1927). Newtons Mechanik und ihr Einflüss auf die Gestaltung die theoretische Physik. *Naturwissenschaften* **15**, 273–6; reprinted in *Ideas and opinions*, pp. 257, 258.
Engel, F. and Stäckel, P. (1895). *Die Theorie der Parallellinien von Euklid bis auf Gauss.* Teubner, Leipzig.
Euclid. *Elements* (transl. ed. T. L. Heath (1956)). Dover, New York.
Forder, H. G. (1927 (reprinted 1958)). *The foundations of Euclidean geometry.* Dover, New York.

Freudenthal, H. (1962). The main trends in the foundations of geometry in the nineteenth century. In *Logic, methodology, and philosophy of science* (eds. E. Nagel, P. Suppes, and A. Tarski). Stanford University Press.
— (1975). Riemann. In *Dictionary of scientific biography*, XI, p. 448. Scribners.
Friedrichs, K. O. (1965). *From Pythagoras to Einstein*. Random House, New York.
Gans, D. (1973). *Non-Euclidean geometry*. Academic Press, New York.
Golos, E. B. (1968). *Foundations of Euclidean and non-Euclidean geometry*. Holt, Rinehart, and Winston. New York.
Gray, J. J. (1979). Non-Euclidean geometry, a reinterpretation. *Hist. Math.*, to be published.
Greenberg, M. J. (1974). *Euclidean and non-Euclidean geometries, development and history*. Freeman, San Francisco.
Hall, A. R. and Hall, M. B. (1962). *Unpublished scientific papers of Sir Isaac Newton*. Cambridge University Press.
Heath, T. L. (1956). *Euclid's 'Elements'*. Dover, New York.
— (1921). *A history of Greek mathematics*. Oxford University Press, Oxford.
— (1930 (reprinted 1963)). *Greek mathematics*. Dover, New York.
— (1949). *Mathematics in Aristotle*. Oxford University Press, Oxford.
Helmholtz, H. (1960). On the origin and significance of the geometrical axioms. In *The world of mathematics* (ed. J. R. Newman). Vol. I, pp. 647–68.
Hilbert, D. (1898–9). *Grundlagen der Geometrie* (English transl. 1902); *Foundations of geometry* (2nd edn.). Open Court, 1971. La Salle, Illinois.
Hilbert, D., and Cohn-Vossen, S. (1952). *Geometry and the imagination*. Chelsea, New York.
Holton, G. (1973). *Thematic origins of scientific thought, Kepler to Einstein*. Harvard University Press, Cambridge, Mass.
Jammer, M. (1969). *Concepts of space*. Harvard University Press, Cambridge, Mass.
Kant, I. (1787). *Critique of pure reason* (transl. N. Kemp Smith (1929)). MacMillan, London.
— *Briefwechsel* (ed. O. Schöndörffer (1972)). Hamburg.
Kilmister, C. W. (1973). *Special relativity, selections*. Pergamon Press, Oxford.
— (1973). *General relativity, selections* (including Riemann's *Hypotheses* (transl. W. K. Clifford). Pergamon Press, Oxford.
Klein, F. (1927). *Vorlesungen über nicht-Euklidische Geometrie*. Chelsea, New York.
— (1926, 1927 (reprinted 1967)). *Vorlesungen über die Entwicklung der Mathematik im 19 Jahrhundert*. Chelsea, New York.
— (1939). *Elementary mathematics from an advanced standpoint—geometry* (transl. E. R. Hedrick and C. A. Noble). Macmillan, London.
Kline, M. J. (1972). *Mathematical thought from ancient to modern times*. Oxford University Press, London.
Knorr, W. R. (1975). *The evolution of the Euclidean elements*, Synthese Historical Library. Reidel, Dordrecht.
Körner, S. (1971). *The philosophy of mathematics*. Hutchinson, London.
Kulczycki, S. (1961). *Non-Euclidean geometry*. Pergamon Press, New York.
Lakatos, I. (1976). *Proofs and refutations*. Cambridge University Press, Cambridge.
Lambert, J. H. (1786). *Theorie der Parallellinien*. In Engel and Stäckel 1895.
Lanczos, C. (1970). *Space through the ages*. Academic Press, London.
Lie, S. (1880). *Math. Annln* **16**, 441–528; transl. by M. Ackermann in R. Hermann

(1975). *Sophus Lie's 1880 transformation group paper*. Mathematical Science Press, Brookline, Mass.
Lobachevskii, N. (1840). *Geometrical researches in the theory of parallels* (transl. G. B. Halsted); see Bonola 1912.
— (1899). *Zwei geometrische Abhandlungen* (1829, 1835) (transl. Scholvin; ed. F. Engel). Teubner, Leipzig.
Lucas, J. R. (1973). *A treatise on time and space*. Methuen, London.
Manning, K. R. (1975). The emergence of the Weierstrassian approach to complex analysis. *Arch. His. Exact Sci.* **14.4**, 297–383.
May, K. O. (1972). Gauss. In *Dictionary of Scientific Biography*, V, pp. 298–315. Scribners, New York.
Meschkowski, H. (1964). *Non-Euclidean geometry* (transl. A. Shenitzer). Academic Press, New York.
— (1965). *Evolution of mathematical thought*. Holden-Day, New York.
Minding, H. F. (1839). Wie sich entscheiden lässt . . . *Crelle* 370–87.
— (1840). Beiträge . . . *Crelle* 323–7.
Minkowski, H. Space and time. in Einstein 1923.
Neugebauer, O. (1969). *The exact sciences in antiquity*. Dover, New York.
Oxford classical dictionary. Oxford University Press, Oxford.
Pedoe, D. (1973). *The gentle art of mathematics*. Penguin, London.
Penrose, R. (1973). Black holes. In *Cosmology now*. BBC Publications, London.
Peters, W. S. (1961). Lamberts Konzeption einer Geometrie auf einer imaginären Kugel. *Kantstudien* **53**, 51–67.
Poincaré, H. (1905). *Essays*. Reprinted 1952, Dover, New York.
Proclus. *A commentary on the first book of Euclid's elements* (transl. G. R. Morrow (1970)). Princeton University Press, Princeton, N.J.
Putnam, H. (1974). *Mathematics, matter and method, philosophical papers*, Vol. 1. Cambridge University Press, Cambridge.
Reichardt, H. (1976). *Gauss und die nicht-Euklidische Geometrie*. Teubner, Leipzig.
Reichenbach, H. (1958). *The philosophy of space and time*. Dover, New York.
Richards, J. (1977). The evolution of empiricism, Hermann von Helmholtz and the foundations of geometry. *Br. J. Phil. Sci.* **28** 235–53.
— (1978). The reception of a mathematical theory, non-Euclidean geometry in England 1868–1883. To be published.
Saccheri, G. (1920). *Euclides vindicatus* (transl. G. B. Halsted (1920)). Open Court, Chicago.
Sambursky, S. (1956). *The physical world of the Greeks*. Routledge and Kegan Paul, London.
Sarton, G. (1959). *A history of science*. Oxford University Press, Oxford.
Schaffner, K. (1972). *Nineteenth century aether theories*, p. 103. Pergamon Press Oxford.
Shirokov, P. (1964). *Sketch of the foundations of non-Euclidean geometry*. Noordhoff, Kasan.
Sommerville, D. M. Y. (1970). *Bibliography of non-Euclidean geometry*. Chelsea, New York.
Spivak, M. (1970). *Differential geometry*, Vol. 2. Includes Riemann *Hypotheses*. Publish or Perish, Kensington, Calif.
Stäckel, P. (1913). *Wolfgang und Johann Bolyai*. 2 vols.
Struik, D. J. (1948). *A concise history of mathematics*. Dover, New York.

— (1969). *A source-book in mathematics, 1200–1800.* Harvard University Press, Cambridge, Mass.
— (1961). *Lectures in differential geometry*, p. 133.
Taylor, E. F. and Wheeler, J. A. (1963). *Space–time physics.* Freeman, San Francisco.
Tilling, L., and Gray, J. J. (1978). J. H. Lambert, mathematician and scientist. *Hist. Math.*, **5.1**, 13–41.
Toth, I. (1967). Das Parallelenproblem im Corpus Aristotelicum. *Arch. Hist. exact Sci.* **3** (4, 5), 249–422.
— (1969). *Non-Euclidean geometry before Euclid. Sci. Am.*
— (Feb. 1977). La revolution non-Euclidienne. *La Recherche*, **75**, 143–51.
Vitale, G. (1680). *Euclides Restituto.* Rome.
van der Waerden, B. L. (1971). *Science awakening.*
Weyl, H (1921 (reprinted 1952)). *Space–time–matter.* Dover, New York.
— (1963). *Philosophy of mathematics and natural science.* Atheneum.
Wolfe, H. (1945). *Introduction to non-Euclidean geometry.* Holt, Rinehart, and Winston, New York.

Index

Absolute geometry, 98
Absolute length, 65
Aganis, 41
Alembert, J. D', 69
Al-Nirizi, 42
Angles, 11, 28
Angle sum, 55
  and area, 66–8
Applications of areas, 25
Arago, F., 165, 167
Aristotle, 1, 19, 31, 39
Asymptotic lines, 61, 79, 101

Babylonians, 1, 14, 23
Beltrami, E., 116, 135–7, 213
Black hole, 207–10
BM 13901, 2
Bolyai, J., 97–9, 107–10, 111, 112, 139, 157
Bolyai, W. F., 80
Borelli, 51

Cardano, G., 36
Cataldi, 51
Ceva, G., 54
Ceva, T., 54
Clavius, 51
Clifford, W. K., 210–11
Codazzi, D., 137
Commandino, 51
Common perpendicular, 59–61
Congruence, 3, 29, 158
Conventions in measurement, 197–201
Corresponding points, 80
Curvature of a curve, 118
  of a surface, 122–4
Curve, 117

Distance between lines, 39
Dostoevsky, F., 161

Einstein, A., 162–3, 166, 170, 172, 174, 191, 196, 205
Elliptic geometry, 142
Equidistant curve, 42
Euclid's *Elements*, 18, 19, 22, 30–2
Euclidean algorithm, 27
Eudemus, 1, 11
Eudoxus, 22
Euler, L., 82–3, 118, 119–20, 168

Figured numbers, 6
Finitude of lines, 64, 142, 150
Fizeau, A., 167

Fresnel, H., 167, 173
Fourier, J., 69

Galileo, G., 128
Gauss, C. F., 68, 76–81, 87, 117, 127
geodesic, 124, 203
gnomon, 6
gravity, 196–7, 204

HAA, HOA, HRA, 55
Helmholtz, H. von, 158, 203
Herodotus, 2
Heron, 128
Hilbert, D., 138, 139, 143
Hippocrates of Chios, 19
Horocycle (=L), 100, 152, 215
Horosphere (=F), 101, 153, 215
Huyghens, C., 125.

Interval, invariance of, 184, 189

Kaestner, A. G., 64, 74, 77
Kant, I., 64, 71
Kennedy–Thorndike experiment, 173
Kepler, J., 163
Khayyam, O., 42
Klein, F., 142, 158, 215
Klügel, G. S., 64

Lagrange, J.-L., 69
Lambert, J. H., 64–7, 71, 77, 84, 87, 109, 156
Laplace, P. S., 69
Legendre, A. M., 69–71, 93, 152
Lie, S., 158
Light clock, 180
Light cone, 179
Lipschitz, R., 161
Lobachevskii, N. I., 71, 96, 99–107, 111, 112, 133, 139, 157
Lorentz, H. A., 167, 170, 173

Maxwell, J. C., 166
Meusnier, G., 120
Michelson-Morley, 167
Minding, H. F., 125
Minkowski, H., 171
Monkey saddle, 127
Monge, G., 118

Nasir Eddin, 42–4, 52
Newton, I., 161, 163–4
Non-Euclidean geometry, 96–116, 144–7

Pappus, 16
Parallel lines, 11, 77
  in non-Euclidean geometry, 79
Parallel postulate, 31–4
Penrose, R., 209
Pfaff, J. F., 74
Plates, heated, 201
  cooled, 202
Plato, 1, 4, 15, 19
Playfair, 37
Plimpton, 332, 13
Poincaré, 170, 205
Poincaré model, 144–7
Points not on a circle, 80
Poncelet, J. V., 69
Prism theorem, 99
Proclus, 3, 4, 36, 38–40, 51
Proper time, 186
Pythagoras theorem, 5, 6–11, 20–1, 181
Pythagoreans, 4–11, 19
Pythagorean triple, 8, 11.

Regular numbers, 14
Rhetorical algebra, 2
Riemann, B., 117, 129–33, 157
Root two irrational, 9, 10

Saccheri, Gerolamo, 54–62, 64, 93, 148–50, 156

Saville, Sir Henry, 54
Schweikart, F. K., 87
Side and diameter numbers, 25
Similarity, 19
Simultaneity, relativity of, 166
Solid geometry, 45–6
Space-time, 176
Special relativity, postulates, 172
Spherical geometry, 63
Squaring the circle, 109
Surface, 119

Taurinus, F. A., 87, 88–91, 157
Thales, 3
Theaetetus, 18
Theodorus, 18
Translation, 29
Trigonometry, 82
  hyperbolic, 84, 92, 188
  spherical, 46, 92, 113–16
Truth in mathematics, 12, 137–8
Twin paradox, 190

Velocity parameter, 188
Vitale, Giordano, 52

Wallis, J., 51